The Politics of Agrarian Change in Asia and Latin America

The Politics of Agrarian Change in Asia and Latin America

Edited by Howard Handelman

Published in Association with the American Universities Field Staff

INDIANA UNIVERSITY PRESS

Bloomington

Manufactured in the United States of America

Library of Congress Cataloging in Publication Data
Main entry under title:

The Politics of agrarian change in Asia and Latin
America

"Originally country study papers presented to the second project seminar, held at the AUFS Institute of World Affairs Conference Center in Salisbury, Connecticut, September 16-18, 1979"—Pref.

"Published in association with the American Universities Field Staff."

Includes bibliographical references.

Contents: An Indian farm lobby, the Kisan Sammelan/Marcus Franda—The farmers federation of Thailand/Brewster Grace—Rural mobilization for modernization in South Korea/Albert Ravenholt—Ecuadorian agrarian reform: the politics of limited change/Howard Handelman—Food policy decision-making in Colombia/Thomas G. Sanders—Peasants, landlords and bureaucrats: the politics of agrarian reform in Peru/Howard Handelman—Agrarian reform and the international consensus/Alan W. Horton

1. Land reform—Asia—Case studies. 2. Land reform—Latin America—Case studies. 3. Agricultural price supports—Asia——Case studies. 4. Agricultural price supports—Latin America—Case studies. I. Handelman, Howard, 1943- II. American Universities Field Staff.

HD1333.A78P64 338.1'85 81-47565
ISBN 0-253-34548-0 AACR2

1 2 3 4 5 85 84 83 82 81

Contents

Contributors

Marcus Franda, American Universities Field Staff Associate for India and Bangladesh since 1971, is the author of *Radical Politics in South Asia* (with Paul Brass) and *India's Rural Development* as well as numerous other books and articles on both the Indian Ocean region and the Subcontinent.

Brewster Grace, American Universities Field Staff Consulting Associate for Southeast Asia, is co-director (with Anne Grace) of the Global Interdependence Center in Philadelphia.

Howard Handelman, American Universities Field Staff Faculty Associate for Latin American 1978-1980, is a Professor of Political Science at the University of Wisconsin-Milwaukee, and director of the AUFS Politics of Food Project 1979-1981. His most recent book (with Thomas G. Sanders) is *Military Government and the Movement Toward Democracy in South America.*

Alan W. Horton, American Universities Field Staff Associate for the Middle East, is the former Executive Director of AUFS (1968-1978) and current Director of the AUFS Center for Mediterranean Studies in Rome.

Albert W. Ravenholt, American Universities Field Staff Associate for East and Southeast Asia, is a specialist on tropical agriculture. He is the author of *The Philippines: a Young Republic on the Move*, and chapters in other volumes as well as numerous *AUFS Reports.*

Thomas G. Sanders, American Universities Field Staff Associate for Latin America, is the author of articles on religion, population, education, and politics in the Southern Hemisphere. His books include *Protestant Concepts of Church and State, Catholic Innovation in a Changing Latin America*, and *Secular Consciousness and National Conscience.*

Preface

This volume is the second to result from study and analysis conducted under the "Politics of Food" project of the American Universities Field Staff.* The project, begun in 1977, has examined key factors affecting decisions about food production and distribution in a series of developing countries, as well as the ways in which international pressures and agencies influence those decisions. With support from the Rockefeller Brothers Fund, the Rockefeller Foundation, and the Tinker Foundation, a series of case studies have been carried out and the resulting papers discussed at seminars.

The American Universities Field Staff (AUFS) is a consortium of educational institutions in the United States that supports a career corps of scholarly correspondents overseas, known as Associates. The latter, working primarily in Asia, Africa, and Latin America, study, evaluate, and report in depth on current developments in their countries and regions, as well as examine selected crosscultural or global themes. Periodically the Associates return to the United States to visit the consortium universities, where they teach classes and consult with scholars and students.

Participants in the Politics of Food project are scholars and food specialists from the countries under review, interested faculty members from universities in the AUFS consortium, and AUFS Associates with a particular concern for issues of agricultural policy.

The project's goal is to shed light on the complex issues involved in food policy-making at the local, national, and international levels and to stimulate wide-ranging discussion in the developing countries, the industrialized countries, and the community of international agencies of the factors and relationships that need to be considered in future decisions concerning food and of the values that various policy choices entail. Project participants have treated agronomic, technological, or narrowly economic issues only in passing because of their belief that some of the chief difficulties in assuring adequate nourishment for the populations of Asia, Africa, and Latin America lie in political and social constraints. Yet these factors are poorly understood and little studied, either in the United States or in the countries themselves.

At the same time, participants propound no simple formulas and, writing from long-time experience in and sympathy with developing countries, they are aware of the enormous difficulties political and intellectual leaders there face and of the necessity to choose courses of action, not in acquiescence to external advice or norms but in accord with autonomous determinations of what is in the best interests of their nations and peoples.

Chapters in this book were originally country study papers presented to the second project seminar, held at the AUFS Institute of World Affairs Conference Center in Salisbury, Connecticut, September 16-18, 1979. Travel grants from the Agricultural Development Council assisted several specialists from Asia to attend the seminar. Revised papers taking account of the seminar discussions were published during 1980 as *AUFS Reports*. Participants in the seminar were:

*The first was Barbara Huddleston and Jon McLin, eds., *Political Investments in Food Production* (Bloomington, Ind. and London: Indiana University Press, 1979).

Paper-Writers and Commentators

Ronald Aqua
Social Science Research Council

José María Caballero
Instituto de Estudios Peruanos
Lima, Peru

Michael Fleet
Marquette University

Marcus Franda
American Universities Field Staff

Howard Handelman
American Universities Field Staff

Alan Horton
American Universities Field Staff

Barbara Huddleston
International Food Policy Research Institute
Washington, D.C.

Brennon Jones
Bread for the World

Syed Karim
Cargill Company

Nelson Kasfir
Dartmouth College

Albert Ravenholt
American Universities Field Staff

Sylvester Rowe
Information Department, United Nations

Thomas G. Sanders
American Universities Field Staff

Bhabani Sen Gupta
Nehru University
New Delhi, India

Raghbir Singh
Punjab Agricultural University
Ludhiana, India

John Thompson (Chairman)
American Universities Field Staff

Other Participants

Charles Carreras
Ramapo College

Betty Hanson
University of Connecticut

Charles K. Mann
Rockefeller Foundation

Peter B. Martin
American Universities Field Staff

Louis Pierro
University of Connecticut

Shlomo Reutlinger
The World Bank

Stanley Seaver
University of Connecticut

Shivaji Sen Gupta
New York

Andrea Simon
Ramapo College

Manon Spitzer
American Universities Field Staff

Grafton Trout
Indiana University

Fred Welz
U.S. Department of Agriculture

James Caplinger
Charles F. Kettering Foundation

Introduction

Howard Handelman

Writing in a 1976 *Scientific American* edition on "Food and Agriculture," Rockefeller Foundation Vice President Sterling Wortman noted: "The world food situation is serious, even precarious. It is also true that the world may have, for the first time in history, the ability to deal effectively with the interacting problems of food production, rapid population growth and poverty.... Yet year in and year out much of the world cannot feed itself...."[1] That statement, coming some two years after the Rome World Food Conference, reflected the growing specter of massive starvation in the African Sahel and other regions. Today—excepting war-torn areas like Kampuchea, Uganda, and Ethiopia-Somalia—the danger of widespread famine appears to have abated, but in Africa, Asia, and Latin America, some one billion people still suffer from malnutrition.

As poverty and hunger in the developing areas have impinged on the world's consciousness (and, perhaps, conscience) there has been an outpouring of social science, technical and journalistic publications on Third-World agriculture and rural development. Mounting evidence strongly suggests that stimulating food production in the less developed countries and, more important, improving the nutritional levels of the poor (two related, but distinct, goals) require complex strategies and solutions that transcend the spread of modern technology and greater agricultural inputs. The structure of Third- and Fourth-World rural society—high degrees of land tenancy and landlessness, extreme inequities in economic and political power relations—often serves as a barrier to increased agricultural output in the face of greater technological possibilities for change. Moreover, as Keith Griffin has eloquently argued, even when such barriers to production are overcome, economic "growth provides no guarantee that the standard of living of the poor [will] improve."[2] Higher food output will not reduce malnutrition and hunger if large portions of the peasantry lack sufficient land to share in that growing production while the urban and landless rural poor lack the purchasing power to buy it.

What this suggests, then, is that raising both food production and consumption usually requires not only economic growth and greater productivity but fundamental political change as well. Presumably, virtually all of Asia's and Latin America's major decision-makers and political participants—politicians, bureaucrats, industrialists, merchants and commercial interests, urban workers, rural landlords, peasants and agricultural workers—favor improving food output and nutritional levels. However, the politics of agrarian change and development frequently pit these groups against each other in what they perceive to be (and often actually is) a "zero sum game"—an economic-political conflict in which one group's benefits come at the expense of the other.

This volume presents six Asian and Latin American case studies analyzing the interplay of rural interest groups, government bureaucrats, and other national (or international) agencies and conferences on the need for agrarian reform and the ongoing debate regarding its nature. The studies of India, South Korea, Thailand, Colombia, Ecuador, and Peru suggest that on some issues—most notably farmgate crop prices—all agricultural producers tend to support a common policy in opposition to urban consumers. In other areas—land redistribution legislation and rural tenancy contracts—peasants and landlords may clash, while urban sectors either sit on the sidelines or actively support one group. Taken together, these case studies offer a tentative conceptual framework for analyzing government agrarian policy in areas such as land reform, crop pricing, and rural modernization programs in the developing world.

Agrarian Reform: Equity and Production

As Alan Horton notes, ever since the 1950s international agencies and conferences dealing with the problem of rural development in the Third World have touched on the issue of agrarian reform with varying degrees of emphasis. Essentially, the term agrarian reform will be used here primarily to mean land redistribution to needy peasantry and the provision of requisite services and credits.[3] In some Asian nations—notably India and Thailand—it may also involve the revision of landlord-tenant rental arrangements. Initially, proponents of such reforms defended them on the basis of social justice, economic equity, and political tranquility. Critics of reform, or those who wished to slow its pace, often argued that equity considerations were in conflict with the need for increasing agricultural production. In Latin America, as rural-to-urban migration swelled the population of the cities (ironically, a migration partly spurred by the lack of agrarian reform in the countryside), conservative landowners and allied political interests argued that their nations "could not afford" agrarian reform because it would interfere with the production of food for the mushrooming urban population. Underlying this argument was the assertion that peasant smallholders lacked either the capital inputs or the technological know-how (or both) to produce as efficiently as larger—presumably more mechanized—landlords who could take advantage of economies of scale.

A growing body of evidence, however, casts doubt on that assumption. All too often large landowners in Asia and Latin America have managed their land inefficiently, have left large portions of it unused or grossly misused (particularly in the *latifundia* of Latin America) or have cultivated crops that do not contribute to the nutrition of much of the population. Marcus Franda notes that in Indian states such as Punjab and Haryana—where land is more equitably distributed—productivity is much higher than in the landlord-dominated states of Bihar and Bengal. Indeed, data from India, Brazil, Bolivia, Ecuador, South Korea, Taiwan, Egypt, and other Asian and Latin American countries consistently show that there is a strong negative relationship (often linear in nature) between farm size and productivity per acre. Thus, James Grant, former President of the Overseas Development Council, has noted, "with other factors equal, small farms are more productive per acre than big farms because the small farmer devotes more hours and care in coaxing maximum production from his limited acreage...." Consequently, adds Grant, "a land and capital scarce (but population-plentiful) country should favor 40 two and a half acre farms over a single-owner 100-acre farm in order to make optimum use of available land, labor and capital."[4]

Albert Ravenholt's study of South Korea—as well as the records of Taiwan and Japan—amply support that position. With a maximum allowable farm size of merely three hectares, South Korea has achieved self-sufficiency in rice production, increased food output consistently faster than population growth, and moved dramatically toward more equitable distribution of income. Other Far Eastern nations have also clearly demonstrated that agrarian reform and equity in landholding are very compatible with high productivity—Japan, South Korea, and Taiwan have among the highest per acre rice yields in the world, at levels of productivity some three times higher than the Philippines where land ownership is far more concentrated. On the other hand, Peru's current experience in the early years of its extensive land reform suggests that the mere transfer of land tenure from *latifundistas* (large owners) to peasants and plantation workers is certainly no guarantee of greater productivity. Indeed without adequate crop prices, credits, and technical assistance, Peru's reformed agrarian sector has stagnated—a common pattern in other, less extensive, Latin American land reforms where support programs were also lacking.

Land Tenancy and Agrarian Reform in Asia and Latin America: The Historical Context

By concentrating on Asia and Latin America, this book has focused on the two regions of the developing world where agrarian reform has been most intensely discussed and debated. Asia, of course, contains the majority of the world's rural population, most of the landless or land-short peasants, and most of the world's malnourished people.[5] The pressure of population on cultivable land is generally much higher in Asia than in Latin America (or Africa) and the proportion of malnutrition greater. Thus, one study estimates that 63 percent of all Asians consume below the requisite daily number of calories as compared to only 36 percent of Latin Americans (and 61% of Africans).[6]

If Asia suffers from more intense rural poverty and general malnutrition, Latin America is characterized by greater inequality in both income distribution and land tenure. Indeed, World Bank figures show that Latin America—where the richest 20 percent of the population typically earns 60-65 percent of the national income—has the most inequitable distribution of income among any of the regions of the world.[7] As Tables 1 and 2 reveal, the traditional pattern of land tenure has also been far more concentrated in Latin America than in Asia (or most of Africa). This has been true prior to and after agrarian reforms of varying dimensions in both regions.

Table 1 reveals the tremendous pre-reform dominance of Latin American agrarian lands by massive latifundia of over 1,000 hectares (2,470 acres)—an unimaginable size in most of Asia. These giant estates—accounting for 1.3 percent of all farm units (less in Peru)—controlled 69-92 percent of private farmland in pre-reform years. Only in Peru and in Cuba has that pattern been dramatically altered by reform legislation. In contemporary Bolivia, Colombia, Ecuador, and Venezuela, less than 2 percent of the farm units still control 50-70 percent of the land subsequent to reform laws. In "postrevolutionary" Mexico, 2.3 percent of the owners (with units of over 100 hectares) now control 47 percent of agricultural property.

Unfortunately, Table 2 does not provide Asian land tenure data prior to World War II. Even before the passage of any agrarian reform laws, however, the region never had large estates of a magnitude approaching the Latin American latifundia and had nowhere near the degree of land concentration.[8] Even among those Asian countries with the greatest concentration of land in 1960—India, Pakistan, and the Philippines—none approached the geni index of Ecuador, Colombia, Venezuela, Bolivia, or Mexico.

One should not, however, exaggerate the degree of rural equity which existed in traditional (prewar) Asia or currently in the Philippines, Thailand, and India. While large estates may never have been dominant or pervasive in countries such as Korea, Ravenholt points to the very exploitative landlord-tenant relationships that frequently existed there. The same was true of prerevolutionary China, the more feudal states of India (particularly prior to agrarian reform), and much of Thailand today. Finally, the land distribution tables presented above merely reflect

Table 1
The Concentration of Land Ownership in Latin America: Before Agrarian Reform

	Size of Units							
	Under 5 Hectares		5-10 Hectares		100-1,000 Hectares		Over 1,000 Hectares	
	% of Units	% of Area	% of Units	% of Area	% of Units	% of Area	% of Units	% of Area
Bolivia	51	0.2	22	1.4	4	6	3	92
Ecuador	73	7.2	25	28.4	1.8	27	0.2	37.2
Colombia	76.5[a]	8.8[a]	16.6[b]	15.4[b]	6.9[c]	75.8[c]	d	d
Mexico	59	0.8	32	5	7	12	2	82
Peru	83	6	16	10	1	14	0.2	69
Venezuela	54	1.3	41	7	4	13	1.5	79

a. under 10 hectares.
b. 10-50 hectares
c. over 50 hectares.
d. included in previous column.

Dates of Data: Bolivia - 1950; Ecuador - 1954; Colombia - 1961; Mexico - 1923; Peru - 1961; Venezuela - 1960.

Sources: Schlomo Eckstein et al., *Land Reform in Latin America: Bolivia, Chile, Mexico, Peru, Venezuela* (Washington: World Bank, 1978); Ministerio de Agricultura, *Evaluación de la Reforma Agraria Ecuatoriana* (Quito, 1978); Salomon Kalmanovitz, *Desarrollo de la Agricultura en Colombia* (Bogota: Carretas, 1978).

Table 2
The Concentration of Land Ownership in Asia (1960)

	Under 1 Hectare		1-5 Hectares		5-20 Hectares		20-50 Hectares[a]	
	% of Units	% of Area	% of Units	% of Area	% of Units	% of Area	% of Units	% of Area
Japan	64.4	30.6	30.6	53.3	5.3	13.1	0.1	3.0
Taiwan[b]	66.4	33.1	32.9	55.7	0.7	11.7		
Indonesia	70.1	28.7	27.4	47.6	2.3	18.1	0.2	5.6
India	60.1	7.6	33.1	46.9	6.2	34.4	0.6	11.1
Pakistan	43.3	7.4	44.6	39.4	12.1	53.2		
Philippines	11.5	1.6	69.5	41.5	18.0	39.0	1.0	17.9[c]
S. Korea[d]	63.3		35.3		1.4		e	
Thailand[f]	18.6	2.5	35.4	26.0	30.7	58.9	1.1	7.9

a. Only the Philippines permitted holdings of more than 50 hectares.
b. Taiwan permitted no holding larger than 10 hectares.
c. 20 hectares or more. Less than 0.25% of the Philippine units were over 50 hectares but they controlled 12.2% of the agricultural area.
d. 1938 (*all* of Korea); By 1960 all South Korean farms were under 3 hectares.
e. Less than .01% of all farms (457 farms of over 2.8 million Korean farms).
f. 1963 data.

Source: D. P. Chaudhri, "New Technologies and Income Distribution in Agriculture," in D. Lehmann (ed.), *Peasants Landlords and Governments: Agrarian Reform in the Third World* (New York: Holmes & Meier, 1974), p. 173. On Korea, see Sidney Klein, *The Pattern of Land Tenure Reform in East Asia After World War II* (New York: Bookman, 1958), p. 245.

the status of those who own some land. Thus, they fail to reflect the large number of landless who constitute 25 percent or more of the agrarian population of India and, more recently, a growing number in Thailand. In short, in all the countries discussed in this book—and in virtually all of Asia and Latin America—traditional tenancy relationships and the rural economic and political power structure cried out for reform.

Starting in South Korea (1948) through Peru (1969), all six of the nations examined here did pass agrarian reform legislation of one kind or another. In two countries—South Korea and Peru—reform has brought about a profound redistribution of land tenancy (with differing degrees of success in production and in improving rural living standards). In three others—Colombia, Thailand, and Ecuador—reductions in landlord dominance (at least through legislation) have been quite limited. The Indian case falls somewhere between those two extremes, with considerable variance from state to state.

What accounts for the varying levels of agrarian reform and rural change in these six nations? To examine the political basis of agrarian change, it is useful to focus on three major factors: first, the extent and nature of peasant political mobilization; second, the landed elite's political and economic power and their adaptability to modernization; third, the role of nonrural ("external") actors, most notably the state, urban power groups, and foreign (transnational) forces.

Modernization and Peasant Political Mobilization

Scholars such as Eric Wolf have argued persuasively that the sweep of Western-style modernization, capitalism, and commercial agriculture into Asia and Latin America and the intrusion of these forces into traditional village society have been major precipitators of peasant unrest in the twentieth century.[9] Without romanticizing precapitalist rural life or minimizing the injustices and exploitation of the old order, one can note certain protective mechanisms in traditional social and economic relationships. These mitigating aspects were often swept aside by the modernization process. Intravillage support systems, reciprocal obligations, and restraints on rural landlords frequently collapsed under the weight of the market system.[10]

As Brewster Grace notes, the origins of recent peasant unrest in Thailand can be traced to the country's incorporation into the world rice trade during the late nineteenth century and to the subsequent rise of middlemen, bureaucrats, and commercial landlords. "The country changed dramatically," Grace observes, "from feudalistic Siam to capitalist Thailand." Whereas under the old order "the Thai monarch tried to restrain oppressive nobles," such restraints were removed by the new military-bureaucratic ruling class.[11] Elsewhere in Asia and Latin America (China, Vietnam, Mexico), where the world economy was subjecting village life to similar shocks, peasants who had formerly accepted the established order rose in far more revolutionary protest.

As commercialization of agriculture made land a more valuable commodity and weakened traditional peasant tenancy rights, peasants and landlords frequently came into conflict. The case studies of India and Peru demonstrate that rural cleavages may be more complex, however, with the peasantry sharply divided between middle peasants (on the margin of becoming small landlords), smallholders, and the landless. In Thailand, the peasantry's discontent over the terms of capitalist agriculture was directed as much toward the merchants and bureaucrats as against the landlords.

In some Asian and Latin American societies—China, Vietnam, Cuba, Mexico—peasant mobilization over the issues of land and the agrarian economy was a root cause of major revolutions. In other cases—Peru, Chile (prior to 1973), and several Indian states—peasant politicization and organization has stopped far short of revolutionary levels, but has exerted important pressure for agrarian reform. On the other hand, several Far Eastern nations (South Korea, Taiwan, and Japan) have implemented successful agrarian reform programs in the absence of significant peasant pressures, due largely to the prodding of the United States or to other external factors such as the North Korean invasion of the South. Finally, in countries such as Colombia, Ecuador, and Thailand and in the more feudalistic regions of India, entrenched rural elites have successfully resisted significant land redistribution.

Of course, virtually every Asian and Latin American nation has been affected by modernization and the growth of the market system, but the extent of peasant mobilization, its intensity, the character of peasant demands, and their success in securing agrarian reform has varied tremendously. A number of crossnational studies suggest that the nature and effectiveness of peasant political activity depends heavily on the internal structure of village society, particularly its level of cohesiveness and cooperation. Recent anthropological and sociological research has challenged earlier assertions that peasant culture is generally highly individualistic and divisive.[12] Indeed, the strong internal cohesion of the Vietnamese hamlet and of Mexican Indian villages in the state of Morales formed the nucleus, respectively, for the Viet Cong and the Zapatista wing of the Mexican Revolution.[13] On the other hand, regional or socioeconomic cleavages within the countryside have weakened peasant movements in Peru and Thailand, while limiting their scope in India.

Other important determinants of the politics of peasant movements are the extent to which the peasantry can form alliances with external groups, the degree to which they receive (or accept) nonpeasant (generally middle-class) leadership, and the ideology of the outside groups or leaders with whom they ally. Peasants led by Fidel Castro, Mao Zedung, or Ho Chi Minh obviously took a different course than those allegiant to Pancho Villa or the Ayatollah Khomeini.

In none of the nations discussed in this volume did rural mobilization take on revolutionary proportions. In Peru, however, large-scale peasant land seizures in the early 1960s helped set the stage for later land reform. The resurgence of the radical Confederation of Peruvian Peasants (CCP) in the early 1970s helped expand the scope of land redistribution and hasten its pace. The highland Indian communities (many with a tradition of communal cooperative projects) served as the base for peasant federations in the 1960s and 1970s. On the other hand, long-standing mistrust between peasants in different regions of the country or conflict between village smallholders and hacienda tenants limited the strength of peasant mobilization.[14] Alliances with urban radicals (in the labor and student movements) had a mixed effect, serving both to sharpen peasant political skills and to heighten government wariness of rural mobilization.

Internal divisions within peasant society have inhibited the rise of broad-based movements in Colombia or of radical ones in India. However, the nature of peasant politics is quite different in those two nations. The growth of commercial agriculture in Colombia has led to periodic eviction of tenants, the consolidation of landholdings in commercial regions such as the Cauca Valley, and the creation of a large, landless rural class that has been forced into the cities or has migrated illegally to neighboring Venezuela. From 1949 through the 1950s these factors contributed to a violent upheaval in the Colombian countryside known simply as "La Violencia." Ultimately, thousands of villagers died in the bloodletting,[15] but that rural violence was anomic and lacking in clearly defined political or class interests. Peasants divided along traditional political party lines (the Liberals and Conservatives, both of which were led by different sectors of the nation's oligarchy) and were more prone to kill other peasants in the opposition party than to attack landlords on their own. Therefore, as Thomas Sanders demonstrates, peasant organizations in Colombia have never been effective forces for agrarian reform or rural change.

India's situation has been somewhat different. Franda describes how the Kisan Sammelan (Farmers' Organization) became a powerful political actor representing the middle peasantry. Indeed, the cultivating castes have become the dominant political force in many of India's northern states and even propelled Kisan Sammelan leader Charan Singh briefly into the nation's prime ministership in 1979. As Franda notes, the spread of political power from the relatively exclusive province of the upper castes (Brahmins, etc.) to middle-ranking castes (Jats, etc.) in the 1970s has led to much "pro-peasant" legislation, including agrarian reform measures, agricultural credits, and better farmgate crop prices (after the Janata administration replaced Indira Gandhi in 1977). However, caste divisions within the rural lower classes have separated the landowning middle peasants as much as from the impoverished Scheduled Castes and the many landless rural laborers (who remain powerless) as from the larger landowners. One fascinating aspect of Indian agrarian change has been the way in which the middle peasant, rather than the larger landlords, has taken advantage of, and prospered by, the technology of the Green Revolution. These benefits, however, have not trickled down to the poorest smallholders or to the landless.[16]

Peasant mobilization in Ecuador has been far more limited and less successful than in either Peru or in India, and even less vocal than in Colombia. The restricted size of the industrial work force (and, hence, of the urban trade union movement) and the absence of organized plantation workers (prevalent in Peru and Cuba) or miners (Peru, Bolivia, Chile) have deprived the peasantry of three important potential sources for the spread of rural mobilization and radicalization. In Ecuador, as in Colombia, whatever land reform has been passed to date has resulted more from extranational pressures exerted by the Alliance for Progress or the United States (as discussed by Alan Horton) than from internal peasant political organization. Similarly, the extent of political mobilization in Thailand has been fairly small, with the short-lived exception of the Farmers' Federation of Thailand discussed by Grace. The Thai villagers' loyalty to Church and King has reduced the likelihood of extensive radicalization, while the localized nature of peasant interests prevented the FFT from addressing issues on a national scale.

Finally, South Korea offers a fascinating case study of a very different kind. Far reaching, and apparently successful, agrarian reform was introduced by the government in the absence of significant peasant pressures. Here, even more so than in Peru's "military revolution," agrarian reform was initiated almost exclusively from above. First under the United States' military occupation and then under the Rhee regime, the Korean authorities carried out full-scale land redistribution, with ownership eventually limited to only three hectares. That reform was followed by an extensive village development program in the 1970s, the Saemaul Undong (New Village Spirit Movement). To be sure, the agrarian reform of the early 1950s was partially a response to the rise of leftist popular village committees and the existence of scattered communist guerrillas after South Korea's 1948 independence, both of which may have induced the government to satisfy the peasantry's strong land hunger. Similarly, the weak rural vote President Park Chung-hee received in the 1971 election probably helped convince him to introduce the Saemaul Undong and to raise crop prices significantly (and, hence, rural living standards).[17] Given the regime's authoritarian control over the country and the omnipresent surveillance of the Korean CIA, however, the government was obviously not forced into rural reform by these very limited peasant pressures. On the other hand, the political weakness of the nation's landowners following the defeat of the Japanese occupation forces, and the external proddings of the United States in the 1948--1954 period, gave the Rhee government a far freer hand (as in Taiwan and in Japan itself) than in most of Asia or Latin America to dissolve large holdings.

In South Korea, the major expansion of commercial agriculture and the thrust toward modernization in the countryside took place *subsequent* to land redistribution. Thus, unlike many other Third-World nations in which mechanization and increased capital inputs have tended to intensify the advantage of large owners over smallholders, in South Korea the benefits of rural modernization have been more equitably distributed. To be sure, the data indicate that the larger a South Korean farmer's plot, the more his family income has benefited from modernization and rising crop prices.[18] Since the limit on land ownership is three hectares, inequality in land tenure and in rural incomes is nevertheless far lower than in most of the Third World. Moreover, unlike Colombia, Ecuador, and Peru—where urban incomes are typically four or five times higher than in rural areas—South Korea is one of the few developing nations in the world where average rural family income equals or exceeds urban levels.[19] However, given the absence of popular participation in Korea's authoritarian political system, there is nothing to guarantee that the government may not decide to lower crop prices once again in the future and thereby reintroduce a rural-urban income gap. Current plans to allow ownership of more than three hectares (to facilitate commercial farming) also may reintroduce some income inequalities.

The Traditional Rural Aristocracy and the Rise of Commercial Agriculture

As Barrington Moore has noted, not only rural politics but the course of an entire nation's political and economic development as well is often greatly influenced by the way in which the traditional rural elite reacts and adapts to the rise of capitalism and commercial agriculture.[20] Thus, for example, the ability of the British gentry to accept the Industrial Revolution and integrate themselves into it (the enclosure movement and the growth of the wool-based textile industry), explains in large part the very different trajectory of modern British history as compared to that of France and Russia. A critical factor for any land-owning class wishing to maintain its economic and political power is its ability to establish and retain legitimacy in the eyes of other socioeconomic groups, not least of which is the peasantry itself. Outbreaks of peasant rebellion and the rise of rural revolutions in Southeast Asia and Latin America have been set off as much by the landlords' loss of "moral legitimacy" as by objective economic or political exploitation.[21] In Latin America—despite the highly inequitable distribution of land—the hacienda-owning class was generally able to contain peasant discontent by establishing a "patron-client" relationship with their tenants, extending certain favors (loans, protection, sponsorship of fiestas) in return for peasant loyalty.[22] This pattern contrasts strongly with the situation in prewar Korea and in prerevolutionary China, where peasant-landlord antipathy was often intense.

A second, closely related, determinant of the landlords' power is their ability to form political alliances with other economically and politically influential groups in society. Does the agrarian elite form alliances with (or intermarry and mix with) urban industrialites and commercial leaders? Or, are these groups distinct, with interests in conflict (over issues such as crop and raw material prices or tariff policy)? Finally, what position do rural landlords occupy in the nation's political power structure? Do they have influential pressure groups that can effectively lobby for their interests? Are they in leadership positions, or at least well-represented in the political party system (if one exists)? Do they have a voice in the nation's press and other mass media?

Colombia presents a classic example of an adaptive, politically sophisticated rural elite which has proven remarkably adept at accepting limited reforms, while deflecting any fundamental challenge to their dominance of agricultural land. Sanders notes the conflicting interests and perceptions that sometimes divide rural and urban industrial elites or landlords and government technocrats. Still, there is sufficient overlap and family interchange among these three sectors in Colombia's highly oligarchical society to insure that the powerful *gremios* (associations of coffee producers, sugar plantation owners, cattlemen, etc.) can exercise near veto power over any legislation that threatens their fundamental interests (while at the same time wisely accepting moderate reformist measures). The successful recent production record of Colombia's modernizing, commercial landlords (most notably in the Cauca Valley) insures this sector a role in Colombia's capitalist development. Concurrently, the influence of the coffee-producing gremios (and associated banking groups) in both the Liberal and Conservative Parties also protects fundamental landed interests. Consequently,

Sanders finds, not surprisingly, that Colombian development policy has stressed bringing unused land into production, creating urban jobs for peasant migrants, and increasing agricultural production. There has been virtually no land redistribution, however, and little or no improvement in the nutritional level of Colombia's poor.

Ecuador's powerful coastal sugar companies and its modernizing highland dairy *hacendados* also proved adept at adjusting to international pressures for agrarian reform (expressed through the Alliance for Progress). Indeed, the commercial dairy ranchers of the north central Andes actively supported the 1964 agrarian reform (and succeeding legislation) because it posed no threat to either their operations or those of other capitalist landowners. It was the more feudalistic landlords of the southern highlands and the inefficient rice and cocoa landowners of the coast (dependent on traditional sharecropping and neofeudal tenant arrangements) who lost whatever land was distributed to the peasantry.

Whereas Colombia and Ecuador have had only very limited land redistribution, in the 1970s Peru's military government carried out one of Latin America's most sweeping agrarian reforms, effectively expropriating all of the nation's large agricultural estates and completely ending the rural aristocracy's erstwhile domination of land ownership. A critical factor in Peru's dramatic transformation was the military's perception that the rural elite (both the traditional highland hacendados and the more capitalistic coastal plantation owners) were a major impediment to industrialization and economic growth.Thus, the once-powerful agrarian aristocracy was stripped of any legitimacy in the eyes of the nation's most potent institutional force.

The Indian and Thai cases are somewhat more complex. In India the landowning classes' political power and their role in land tenancy have varied considerably from state to state. Not surprisingly, Franda notes, the extent of agrarian reform (a jurisdiction of state government) has varied correspondingly. Given India's comparatively recent colonial history (an experience which usually weakens the authority and legitimacy of traditional elites) relative to Latin America and the tendency of the Congress Party to favor the Westernized, urban elite, the agrarian landed aristocracy suffered a loss of political power in many states. However, because the caste system tended to deflect open class conflict and because of the more decentralized nature of Indian politics (which enabled regional power groups to retain influence at the local level), landlords were able to maintain considerable control in the more feudalistic Indian states such as Bihar and Bengal and to block agrarian reform in those regions.

Brewster Grace indicates that in Thailand landlords were also able to prevent meaningful agrarian reform or effective controls on rent by forming political alliances with rice merchants, bureaucrats, village headmen, the rural police, and the military. Since the commercial rice trade was dominated by immigrant Chinese, the merchant group's legitimacy was reduced in the eyes of the peasantry. At the same time, however, perceived exploitation by merchants and moneylenders may have diverted some peasant resentment away from the landlords. Finally, Thailand's proximity to communist Laos and Cambodia/Kampuchea and the military's fear of guerrilla insurrection (particularly in the traditionally rebellion-prone northeast) have apparently induced recent military regimes to take a harder line in resisting grassroots peasant mobilization.

In contrast, South Korea represents the most striking case of a landowning elite which lost all political legitimacy or power. Since Korean farms—even prior to agrarian reform—rarely exceeded 50 acres, landlords historically did not exert as much national economic or political power as the Latin American *latifundistas* or even the regional power of Bihar and Bengal's neofeudal lords. Moreover, Ravenholt notes, a significant portion of the larger and richer landlords during the country's prewar, Japanese occupation were either Japanese "colonialists" or their local allies. As a consequence, the landed elite had no legitimacy after the war either from the perspective of the Korean peasantry or of the officials in the new Syngman Rhee government. With the defeat of the Japanese and the creation of an independent South Korea state, first the Japanese landlords themselves and then their Korean allies—neither of whom had any defenders—were stripped of their land. By 1954, almost all the large agricultural units had been transferred into the hands of their former tenants.

"External" Actors

Up to this point, our analysis has focused on the way in which agrarian policy is influenced by the balance of power in the countryside. Most of the case studies presented here clearly demonstrate that other important political actors, outside the rural setting, also critically influence the politics of agrarian change. Government policy-makers, for example, are *not* merely calculators, weighing the relative political and economic influence of landlords, peasants, and urban interests and then endorsing the policy preferences of the most powerful groups. The analyses of South Korea, Peru, Colombia, and Thailand demonstrate that government bureaucrats and technicians (and other representatives of the state) make decisions that are often independent, in varying degrees, of rural or urban interest-group influence. These officials have ideological and programmatic preferences of their own which may be the critical determinant of state policy in political systems that have low levels of popular participation.

This is most clearly demonstrated in South Korea and Peru. While the agrarian reform programs implemented in those countries were widely endorsed in the countryside, we have already indicated that they were introduced almost exclusively from above, with little initial peasant pressure or input.[23] In both countries military leaders or civilian technocrats had decided that agrarian reform was a necessary ingredient for long-term economic development and sociopolitical stability. Similar factors motivated the Park regime to institute the New Village Spirit Movement in the 1970s. Wishing to head off dangerously high levels of rural-to-urban migration (and perhaps to further stimulate South Korea's industrial growth by broadening the consumer class), the government undertook a village modernization program.[24] But, such "top-down" reform has its costs. During the early days of Korea's Saemaul Undong there was little peasant influence, and local bureaucrats frequently alienated villagers through the use of highly imperious methods. In Peru, village-level administrators for SINAMOS (the popular mobilization movement established by more radical elements of the military government) often acted similarly.

Grace explains that the powerful Thai military and bureaucracy also make decisions independent of pressures from lower-income rural interest groups. Unlike Peru, the Thai state apparatus—particularly during the frequent periods of military rule—has usually backed the rural landlords, granting them a free hand to repress independent peasant movements such as the FFT. In Colombia, Sanders contends that technocrats in the executive branch (particularly the president's office) are more independent of rural elite pressures (from the gremios) than are the national Congress or the major political parties. Thus, Sanders argues that the technocrats have been more committed to progressive change in the countryside and cities than has the elected legislature. However, the absence of meaningful agrarian reform suggests that these technocrats have stressed increased agricultural production, with little concern for improving equity in land tenure or in income distribution.

There is one other important external actor in the agrarian political process whose role merits discussion here. Alan Horton's chapter on the growing "International Consensus" points to the influence of world opinion and of foreign assistance donors—industrialized nations and international agencies. Agrarian reform was a major *domestic* issue as far back as the Mexican and Russian Revolutions (1910, 1917). The case for such reform as a stimulus to political and economic development, Horton notes, was first introduced widely into international discourse only during the 1950s. International support for agrarian reform—at least at the rhetorical level—underlay the United States policy toward Latin America during the Alliance for Progress of the 1960s. Finally, the case for reform gained a dominant position in the proclamations of international agencies and conferences in the 1970s.

In South Korea, Japan, and Taiwan, the United States obviously exercised a critical role in pushing through the initial stages of land redistribution. Indeed, the Far East may well be the only region of the Third World where the U.S.—or any other "foreign actor"—successfully insisted on meaningful land reform. Rhetorical commitments to the contrary, neither the United States nor any international agencies have exercised a similar influence elsewhere in Asia (notably Thailand, precommunist South Vietnam, or the Philippines) or in Latin America.

To be sure, the case studies of Colombia and Ecuador do show that pressures or inducements from the Alliance for Progress (i.e., the U.S.) were critical in getting these two nations to pass agrarian reform legislation in the 1960s. Still the data on land redistribution and tenancy 15 years later show that little change had occurred (at least as the result of government policy). In both nations a small rural elite continued to dominate production and tenancy, suggesting that both the external impulse for reform and the legislation itself were largely symbolic. The same was true of the weak 1964 law in Peru introduced by reformist President Belaúnde. Conversely, Peru's very extensive land redistribution in the 1970s was carried out in the absence of any international pressures.

Undoubtedly, the rhetoric of international conferences and subsequent proclamations are important in and of themselves. More significantly, they often exert a real influence over policy-making, just as domestic political rhetoric may ultimately affect national level policies. However, as Horton notes, within the emerging international consensus on the need for agrarian reform the exact nature of such change is interpreted quite differently by developed and developing countries—as well as by radical and conservative regimes among the less developed countries.

Moreover, only time will tell whether the United States (AID), World Bank, and FAO will go beyond this symbolic rhetoric, and insist that recipient nations carry out agrarian reform and institute domestic programs truly geared toward helping the rural poor as a prerequisite for receiving foreign aid. Unfortunately, some cynicism by observers is understandable here. José María Caballero (a noted authority on Peru's agrarian reform program) wryly notes that the growing international consensus in support of agrarian reform crystalized at the very time when such reform efforts had generally either been completed (South Korea, Peru, India) or been abandoned (Colombia, Thailand). That is to say, in the internal political dynamics of much of the Third World the issue had already become somewhat "passé." Indeed, Sanders notes that in Colombia a former World Bank expert serving as an influential adviser to the Pastrana administration (1970-1974) strongly argued *against* agrarian reform to a more-than-receptive national government. Opting for alternative urban employment programs (or perhaps hoping that many displaced peasants would continue to "disappear" across the boarder to Venezuela), the Pastrana government and its successors have, in effect, endorsed large-scale commercial farming that has impressively increased production (for export and for national consumption) but also has pushed small farmers off the land, thereby expanding rural poverty. In a potentially similar vein, during the late 1970s the Park regime considered legislation loosening South Korea's long-standing, stringent limits on the size of farm holdings so as to permit larger commercial farms.[25] Ecuador, whose newly installed civilian government has committed itself to further agrarian reform, may prove to be an exception to this trend.

The Politics of Crop Pricing

Most of the Asian and Latin American case studies discussed here point to land tenure arrangements—ownership or rental terms—as a core issue in the politics of rural change. Peasant spokesmen along with sympathetic "outsiders" (urban radicals or reformers, international agencies of late) have usually focused on agrarian reform of some type—land redistribution, more equitable rental terms or even colonization of virgin lands—as a sine qua non for progressive rural change. But, as Albert Ravenholt points out so well (and Peru's recent experience so clearly demonstrates), "land reform alone is not enough." Unless smallholding peasants (or peasant cooperatives) are given necessary credits, technical assistance and, most important, "fair" prices for their crops, land redistribution programs will neither significantly improve rural living standards nor increase national food production.

The level of crop prices, then, becomes another central nexus of agrarian politics. In most of Asia and Latin America the state—not the market mechanism—sets farmgate produce prices. Consequently, those prices are manifestly a political, rather than a purely economic, issue. Here—unlike the question of agrarian reform—political debate and conflict presumably find landlord and peasant allied together against urban interests.

In most of the nations discussed here (with the notable recent exceptions of South Korea and India), government policy has been to maintain low crop prices so as to provide cheap food for the urban population, inexpensive raw materials for industry, and competitively priced exports. Commonly, the farmer is paid a price below that prevailing on the world market. In some areas local produce prices are also purposefully depressed through the importation of grain (often that contributed by United States foreign assistance programs—PL-480).[26] The creation of "artificially low" crop prices is based on a number of political considerations. First, as

Grace's study of Thailand suggests, it reflects the growing political and economic power of urban industrialists who desire inexpensive raw materials as well as cheap food for their workers (simultaneously demonstrating the commitment of many governments to industrialization at any cost). Second, it indicates that, in Latin America particularly, blue and white collar workers and the urban middle class are more politically influential than the peasantry (or even the rural elite) on this issue because they are better organized (into unions or professional associations), more likely to vote (metropolitan Lima contained over 40% of Peru's entire electorate in the 1980 election), and closer to the centers of political power. Sanders' study of Colombia shows that that nation's powerful landlords have successfully blocked meaningful agrarian reform but have not been successful in their quest for good produce prices. The same is generally true of Thailand and Ecuador. In Peru, the military's "rural revolution" gave the land to the peasants who till it, but failed to give them decent crop prices lest it arouse the ire of the urban population. There, as elsewhere in the developing world, the specter of urban food riots is a more sobering scenario and a more immediately threatening prospect for government officials than are the long-term effects of rural poverty. In any event, peasants may revolt over the issue of land, but they do not seem to do so over crop prices. Their more likely long-term response is simply to migrate to the cities.

"Cheap food" policies might conceivably be justified as a means of serving the nutritional needs of the urban poor. However, low produce prices are generally deleterious in terms of equity considerations and act as a disincentive to domestic food production. In Latin American nations such as Mexico, Colombia, and Peru (or prerevolutionary Cuba), we have seen that urban incomes tend to be four to six times times greater than those in the countryside. Even when compensations are made for income-in-kind consumed by subsistence farmers, the urban-rural income gap is still very great. Thus, while the rural sector accounts for some 45 percent of Latin America's entire population, it contains perhaps 70 percent of the region's malnourished people.[27] Of course cheap food policies do benefit agrarian workers and the rural landless, but they also effectively subsidize the living standards of more affluent urban workers and the middle class at the expense of impoverished peasant smallholders. Indeed, the exploitation of the Latin American countryside by more developed urban centers is so pronounced that local sociologists speak of a system of "internal colonialism" whereby the rural regions serve as a source of cheap labor, food, and raw materials for the cities.[28] Brewster Grace describes a similar situation in Thailand.

India and South Korea, however, now pursue a far different policy. There (as in Japan), farmers are paid relatively generous prices for their crops, often above world market levels. Indian pricing policy clearly reflects the growing political power of the middle peasantry as described by Franda. Unlike most developing nations, India has a well-established democratic electoral system (despite a brief authoritarian interlude under Indira Gandhi) which has produced governments more responsive to popular pressures than is true of Thailand or most of Latin America. Moreover, since the great majority of voters in India are found in the villages (the reverse of the Latin American situation), rural interests are obviously more effectively represented.

Authoritarian South Korea, of course, currently enjoys no such electoral pressures. Through the late 1960s, Korean farmers were paid artificially low prices, like most of their Asian compatriots. During the past decade, however, the military regime has dramatically raised produce prices in a conscious policy of transferring wealth from urban areas to the countryside. As a result, incomes have risen even faster in the villages than in the cities during South Korea's recent economic boom (the reverse of the pattern in Mexico, Brazil, and Colombia). As we have seen, the government's generous pricing policies and its modernization program in the countryside were partially induced by General Park's reduced rural electoral strength in the 1971 presidential election (prior to his regime's more intense clamp-down on legitimate political opposition). However, grassroots pressures on the dictatorial state were minimal and Park's "pro-rural" policies were basically generated from technocratic (i.e., "nonpolitical") considerations. Government policy was designed to reduce the tide of rural-to-urban migration (which had become worrisome to government planners) and perhaps also to create a broader domestic (rural) market for South Korea's rapidly burgeoning consumer goods industries.

The Indian and South Korean governments have tried to pursue the best of both worlds for their urban and rural constituencies by setting

Table 3
The Political Dynamics of Agrarian Change

	DOMESTIC AND INTERNATIONAL Independent Variables					REGIME IDEOLOGY (Intervening Variables)	RURAL POLICY (Dependent Variables)	
	Degree of Peasant Mobilization	Radicalization of peasantry	Adaptability of Rural Elite	Legitimacy of Rural Elite	International Pressures	Orientation of the State	Degree of Agrarian Reform	Level of Crop Prices
ECUADOR	Low	Low	Mixed[a]	High	Varied[b]	Centrist to Moderate Reformist	Low	Low
THAILAND	Generally Low	Low	Mixed[a]	High	Low	Usually Conservative	Low	Low
COLOMBIA	Generally Low	Generally Low (High in Some Areas)	Generally High	High	Varied[b]	Conservative to Centrist Capitalist Developmentalist	Low	Low
INDIA	High for Middle Peasantry Only	Generally Low	Mixed[a]	Mixed[a]	Low	Reformist (Social Democratic)	Varies Greatly from State to State	Moderately High
PERU	Moderate (High in Some Areas)	High in Some Areas	Varies from Region to Region	Low	Varied[b]	Left-Center Nationalist (Mixed Economy)[d]	High	Low
SOUTH KOREA	Low	Low	Low	Low	Initially High[c]	Capitalist with High State Interventionism (Central Planning)	High	High Since 1970

a. Varies according to region or state (or, occasionally, according to crop grown).
b. There was moderate pressure for reform in the 1960s (Alliance for Progress); Very little international pressure in the 1970s.
c. Agrarian reform was first initiated after World War II by the U.S. military occupation; Subsequently there was little external (international) pressure for rural reform.
d. Characterizes the "revolutionary" military regime of General Velasco Alvarado (1968-1975) which carried out Peru's radical agrarian reform; In 1975, with Velasco's ouster, the military government moved rightward.

relatively high farmgate produce prices and then selling food to at least some consumers at a lower, subsidized price (e.g., Indian ration shops for the poor). Of course, that policy puts a strain on public finances and must ultimately be paid for by the taxpayer.

Conclusions

The foregoing discussion of the politics of agrarian change has analyzed the interaction of key sectors of the population with one another and with external (particularly international) actors—if any—for promoting agrarian reform ("independent variables"). Their ideas are interpreted by, and filtered through, government decision-makers who have ideological and programmatic preferences of their own (intervening variable). The resulting government policy (dependent variables) in areas such as agrarian reform and crop pricing thus reflect both national (or international) pressures and the regime's own ideology. The analytical scheme expressed in Table 3 summarizes these relationships.

Table 3 suggests that peasant political mobilization or international pressures may increase both the likelihood of agrarian reform and of higher produce prices—that is, outcomes more favorable to the peasantry. However, these do not seem to be the most salient factors. Of course, one cannot generalize from only six case studies. Nevertheless, the experiences of these nations highlight the critical importance of two other variables: first, the ability of landowning rural elites to maintain their legitimacy and to forge alliances with the military or with other powerful groups (industrialists, bankers, etc.); second, the ideology and programmatic orientation of the governing regime and of its top policy-makers.

This is not to say that there are no other important influences on rural policy. Peru's extensive agrarian reform program in the 1970s, for example, *was* inspired partially by peasant land seizures and the expansion of politically oriented peasant federations during the previous decade. Similarly, the higher crop prices currently paid to Indian farmers (as set by the state) reflect the increasingly potent political organization of that nation's middle peasantry. On the other hand, South Korea currently has one of the most "pro-peasant" agrarian structures (in terms of land tenure, produce prices, and rural modernization programs) of any nation discussed in this volume. This has emerged from an authoritarian political system with virtually no peasant mobilization and little grassroots influence. The critical factor in both Peru and South Korea was that very extensive land reform was carried out when the larger rural landlords lost their legitimate standing with the state and with other relevant and powerful political actors. It was that loss of standing, coupled with the development-oriented ideologies of both the Velasco military government (left-of-center) in Peru and the Rhee and Park (rightist, authoritarian) Korean regimes that produced far-reaching agrarian reform. Peasant mobilization and even foreign pressures were secondary.

On the other hand, the Colombian landlord class adapted very well to the exigencies of commercial farming, the nation's capitalist development model, and Colombia's growing involvement in the world market economy. The rural aristocracy's close links to the country's industrial, banking, and political elites guaranteed the landowners' survival (and, concomitantly, the death of agrarian reform) in the face of troublesome rural-to-urban migration and international pressures (expressed through the Alliance for Progress) for rural reform.

One last point deserves mention. The six national case studies presented here suggest that, while the ideology and policy preferences of top government leadership are a critical determinant of rural policy, a simple "left to right" categorization may not be particularly useful in explaining government decisions. On the whole, Marxist governments in the developing world (Cuba, China, Vietnam) have tended to pursue policies that redistribute income from the cities to the poorer countryside. Conversely, many nations developing under a more capitalistic model (Brazil, Mexico, Thailand) have favored the cities. Ironically, General Park's conservative regime (like the government of Taiwan) has established a higher level of urban-rural income equity than did General Velasco's left-nationalist government or even the regimes in Cuba and the People's Republic of China. From the peasants' perspective, then, whether a government's economic policies are socialist or capitalist may be less important than whether the state supports urban-centered economic growth or a more balanced development model.

NOTES

1. Sterling Wortman, "Food and Agriculture," *Scientific American* 235 (September 1976), p. 30.

2. Keith Griffin, *Land Concentration and Rural Poverty* (London: MacMillan, 1976), p. 2. Indeed, Griffin cites an extensive statistical analysis by two economists which indicates that, for much of the developing world, capitalist economic growth (expressed in rising GNP) has *worsened* the living standards of the bottom half of the population. See Irma Adelman and Cynthia Taft Morris, *Economic Growth and Social Equity in Develping Coiuntries* (Stanford, Calif.: Stanford University Press, 1973).

3. Some land reform programs (Colombia, Ecuador, Venezuela) have stressed the distribution of previously uncultivated lands. Since such property is usually remote, lacking in infrastructure and often of poorer soil quality, these "colonization programs" serve primarily as a means of avoiding meaningful redistribution of quality lands and rarely do much to alleviate the plight of the landless or land-poor peasantry (though they may help a small number of peasants). Redistributive programs may turn over the land directly to smallholders (South Korea), may turn them into peasant cooperatives with varying degrees of state intervention (Peru and Cuba), or may convert them to state farms (Cuba, China). None of the countries examined in this book have used the third strategy, although Peru's critical coastal sugar plantations (officially worker co-ops) are virtual state farms. The evidence seems to suggest that, at least in the short-run, the first strategy (distribution to private smallholders) will lead to the most immediate gains in production if the smallholders are given adequate state aid in the form of crop prices, technical assistance, credits, training, etc. Results, however, vary greatly from country to country and even from crop to crop.

4. James Grant, "Development: The End of Trickle Down?" *Foreign Policy* 12 (Fall 1973). For further documentation on this point, see the article on Ecuador in this volume. Also, T.J. Byres, "Land Reform, Industrialization and the Market Surplus in India," in David Lehmann, ed., *Peasants, Landlords and Governments: Agrarian Reform in the Third World* (New York: Holmes & Meier, 1974). Grant notes that in India farms of 5 acres or less are 50 percent more productive per acre than are farms of 50 acres or more. See also, Griffin, op. cit.

5. By one estimate, 707 million of the world's 1.07 billion undernourished persons in the mid 1970s lived in Asia as compared to 112 million undernourished in Latin America. See, Marcelo Selowsky, *The Economic Dimensions of Malnutrition of the Young* (Athens: Paper presented at the International Symposium on the Child in the World of Tomorrow, July 1978).

6. Ibid.

7. World Bank, *Redistribution With Growth* (Washington: IBRD, 1974).

8. In prewar Korea, for example, even under Japanese occupation, there were only 457 farms larger than 20 hectares in the whole country and only 5,341 (out of more than 2.8 million farms) with 10-20 hectares. Thus, such "large" Korean farms as existed *before* South Korea's postwar agrarian reform (which limited ownership to 3 hectares) were minuscule compared to the *latifundia* of over 500 or 1,000 hectares that have dominated Latin America's countryside. Data on pre-reform Korea are drawn from Sidney Klein, *The Pattern of Land Tenure Reform in East Asia After World War II* (New York: Bookman Associates, 1958), p. 245.

9. Eric R. Wolf, *Peasant Wars in the Twentieth Century* (New York: Harper and Row, 1969). Wolf's book is an indispensable classic for anyone interested in modern peasant revolutions and in theories of peasant unrest. For a more recent analysis and summary of the literature, see: Joel Migdal, *Peasants, Politics and Revolution in the Twentieth Century* (Princeton, N.J.: Princeton University Press, 1974).

10. For the day in which capitalism disrupted traditional society and reduced peasants to an "economic input" in industrializing Europe and elsewhere, see the seminal work: Karl Polanyi, *The Great Transformation* (Boston, Mass.: Beacon Press, 1957).

11. Quoted from James Scott, *The Moral Economy of the Peasant: Rebellion and Subsistence in Southeast Asia* (New Haven, Conn.: Yale University Press, 1976), p. 54.

12. For examples of very influential earlier scholarship that portrayed peasant society as highly internally divisive and fatalistic, see: Edward Banfield, *The Moral Basis of a Backward Society* (New York: Free Press, 1958); George M. Foster, "Peasant Society and the Image of Limited Good," *American Anthropologist* 67 (April 1965), pp. 293-315.

13. On the revolutionary potential of the traditional peasant village in those two nations, see: John McAlister and Paul Mus, *The Vietnamese and Their Revolution* (New York: Harper and Row, 1970); John Womack, Jr., *Zapata and the Mexican Revolution* (New York: Alfred Knopf, 1969).

14. See, Howard Handelman, *Struggle in the Andes: Peasant Political Mobilization in Peru* (Austin, Tex.: University of Texas Press, 1974); also, the article on Peru in this volume.

15. Germán Guzmán, *La Violencia en Colombia* (Cali, Colombia: Progreso, 1968).

16. The exact economic effects of the Green Revolution (new rice breeds and technology) on different strata of rural India have been hotly debated. Some authors have claimed that because of the capital investment needed to use "green revolution" seeds, the new technology has given the larger farmer an advantage over the capital-

poor smallholder. See D.P. Chaudhri, "New Technologies and Income Distribution in Agriculture," in D. Lehmann, ed., *Peasants, Landlords and Governments* (New York: Holmes & Meier, 1974); F.R. Frankel, *India's Green Revolution: Economic Gains and Political Costs* (Princeton, N.J.: Princeton University Press, 1971). Franda has challenged that analysis in the past. A growing consensus of scholars now supports Franda's contention that it has been the middle peasant, rather than the large landlord, in India that has most effectively gained from green revolution technology. This is due partially to the Indian government's policy of providing very inexpensive tube wells to such peasants for irrigation as well as the middle peasant's intensive labor. Yet, Franda admits that the poorest smallholders may have been placed at a competitive disadvantage relative to the middle peasantry and many of the former may have abandoned their plots. For a more general negative assessment of new technologies: Keith Griffin, *The Political Economy of Agrarian Change.* (Cambridge, Mass.: Harvard University Press, 1974).

17. Young Whan Kihl, "Politics and Agrarian Change in South Korea: Rural Modernization by 'Induced' Mobilization" in Raymond Hopkins et al., *Food, Politics and Agricultural Development* (Boulder, Colo.: Westview Press, 1979), p. 156. General Park received strong rural electoral support in the 1963 and 1967 presidential elections. But, as Ravenholt and Young note, stagnating rural living standards in the late 1960s caused a sharp fall in Park's vote in the countryside in the 1971 election. Indeed, Park was defeated in the Southwest by opposition leader Kim Dai Jung of the New Democratic Movement. The General's weaker 1971 vote caused him to initiate a massive rural development program, raise rural incomes through sharply higher crop prices (see Ravenholt's article in this book), and crack down repressively on electoral opposition groups.

18. Ibid., p. 163. Young notes that South Korean farmers with over 2 hectares of land (6.8% of all farmers) had income gains from 1970-1976 (due to sharply higher produce prices) that were over 3 times as great as the gains registered by farmers with under 1 hectare (65.4% of the farm population). The gap in *total* income (as opposed to income *gains*) is presumably less marked. In any event, rural income gaps in South Korea (which has one of the most equitable distributions of income in the world—approximately seventh most equal of some 75 nations for which the World Bank has data) are minuscule when compared to Latin American nations and much of Asia.

19. *The Korean Economy: Growth, Equity and Structural Change* (Seoul: Economic Planning Board, 1976), p. 16. Rural family incomes matched urban levels from 1963-1965, fell to 70 percent of urban levels in 1967-1970 (a fact reflected in Park's weaker rural electoral support in 1971), and rose to parity with urban incomes in 1974-75. In some years, rural income has exceeded urban levels. There is some debate on these figures since rural family income data often include wages earned by family members who are working temporarily in the city. The details, however, do not affect the underlying fact of high urban-rural equity.

20. Barrington Moore, Jr., *Social Origins of Dictatorship and Democracy: Lord and Peasant in the Making of the Modern World* (Boston, Mass.: Beacon Press, 1966). Moore's work is generally recognized as one of the most widely sweeping and brilliant works of social history to appear in recent decades. It analyzes the rise of capitalism, commercial agriculture, industrialization, liberal democracy, fascism and communism in England, France, Germany, Japan, the United States, China, India, and elsewhere.

21. See Scott, op. cit., for a highly stimulating discussion of this point. Scott's book and other articles by him cited in that volume have shed much light on this point.

22. Note the widespread use, by Latin American peasants, of the word *patrón* to address or describe the hacienda owner, thereby clearly defining the patron-client relationship. Scott and others note that this relationship is reciprocal but not equal with the patron clearly benefiting from what is, for all its ameliorating features, still an exploitative relationship.

23. My article on Peru in this book argues that the reformist military regime's initial decision to carry out agrarian reform *was* influenced by widespread peasant land seizures and mobilization in the early 1960s. Resurgent peasant mobilization in the mid-1970s also expanded the scope and pace of the new land redistribution program. Thus, some peasant pressure was present. At the actual time that the military promulgated its radical agrarian reform law in 1969, however, the countryside had been fairly peaceful for over four years. Thus, experts such as economist José María Caballero (who worked with the agrarian reform in its early years and is one of Peru's leading authorities on the subject) argue that there was no pressure from below when the Generals and Colonels decided to radically restructure the countryside.

24. Young Whan Kihl, op. cit.

25. Ibid.

26. Thus, one Agriculture Department technocrat associated with the "Food for Peace" (PL-480) program indicated to me he felt that, except for emergency famine situations, the program did more harm than good by discouraging domestic farm production in recipient nations. For a discussion of the possible negative and positive effects of PL-480, see: Comptroller General of the United States (Report to the U.S. Congress) *Disincentives to Agricultural Production in Developing Countries* (Washington, D.C., 1975), pp. 25-30; Lana Hall, *The Effects of P.L. 480 Wheat Imports on Latin American Countries* (Ithaca, N.Y.: Cornell University Department of Agricultural Economics, 1980).

27. Government Accounting Office, "World Hunger: Implications for U.S. Policy" (Washington, D.C.: Unpublished paper, 1979).

28. Rodolfo Stavenhagen, "Basic Needs, Peasants and the Strategy for Rural Development," in *Another Development: approaches and strategies*, ed. Marc Nerfin (Uppsala: The Dag Hammarskjold Foundation, 1977).

Indian farmers

An Indian Farm Lobby: The Kisan Sammelan

Marcus Franda

India has an agrarian structure which is quite different from that found in other parts of the world. In general, Indian landholdings tend to be smaller than elsewhere, the number of landless and marginal farmers far greater, and population pressures on the land much more intense. The kind of rural poverty found in India—where more than 200 million people are estimated by the Indian government to be below the *Indian* poverty line—is simply not present in other parts of the world.[1]

At the same time, however, there is a dynamic within India's rural areas that is frequently overlooked by those who forecast only gloom, doom, and starvation. India's foodgrain production has more than tripled since the country received its independence in 1947 (foodgrain output per year has jumped from 40 million tons in 1947 to 130 million tons in 1979). Population growth over the same period has not yet doubled (the size of the population has grown from 330 million to 640 million). Land reforms in India have not succeeded in transferring land to the landless in any significant measure, but they have done away with the old *zamindars* and other feudal lords who held or controlled landholdings measured in the hundreds and thousands of acres.[2] Whereas in most Latin American nations the "big landlords"—anyone with 100 hectares (240 acres)—commonly control 50-60 percent or more of the land, in India a big landholder now owns about 10 hectares and as a category controls only 30 percent of the land (see Table 1).[3]

Figures for landholdings in India as a whole mask a most complex maze of different situations in the 22 states, since land reforms and other agricultural policies are subjects reserved to the states in India's federal constitution. There have been attempts by the federal government to influence the states to undertake nationally devised land reform programs, but central law-making and enforcement have not been possible under the constitution. Each state, therefore, has its own history of land reform legislation and implementation.

In the more prosperous agricultural states—Punjab, Haryana, the western portions of Uttar Pradesh, northern Rajasthan, Gujarat, and pockets of Maharashtra, Tamilnadu, Andhra, and Karnataka—there has been a significant transfer of land rights, and occasionally land itself, to a middle-ranking group of peasant castes who were previously sharecroppers, tenants, or smallholders. There has also been consolidation of holdings on a larger scale than elsewhere, more extensive land improvements, and more effective use of modern agricultural inputs (seeds, fertilizers, pesticides, irrigation, and mechanization). In the poorer agricultural regions—and especially in Bihar, Bengal, Orissa, and northeastern India—older feudal patterns of land ownership have prevented cultivators from consolidating their lands and have provided little incentive for land improvement or effective use of irrigation and other modern inputs.

The most extreme contrast is between Punjab and Haryana on the one hand and Bihar and Bengal on the other.[4] Punjab-Haryana account for less than 3 percent of India's population and less than 4 percent of its total land, and yet the two states grow more than 14 percent of India's food. Bihar-Bengal contain almost 16 percent of India's population and nearly 9 percent of its land, but together grow only 12 percent of the food. The reasons for such discrepancies are complex, but landholding patterns are an important part of the explanation. Whereas more than 70 percent of all rural Punjabis own land

Table 1

Size and Distribution of Operational Holdings in India *
1970-1971 Agricultural Census

Size of Holding	Category of Holding	Percent of Total Number	Percent of Acreage Held
25 + acres	Large Landholder	4%	30%
10 - 25 acres	Medium Landholder	11%	30%
2.5 - 10 acres	Small Landholder	35%	31%
0 - 2.5 acres	Marginal Landholder	50%	9%

*An operational holding includes all land used for agricultural production as part of the same technical and economic unit, regardless of title, size, or location.

Source: India, Ministry of Agriculture and Irrigation, Department of Agriculture, *All India Report on Agricultural Census, 1970-71* (New Delhi: Government of India, 1975), p. 26.

and more than 80 percent are owner-cultivators, rather than tenants, in Bihar and Bengal less than 50 percent of all rural residents have rights to land and most are in tenancy or sharecropping arrangements.

Much of the increase in agricultural production in Punjab-Haryana and other food surplus areas is related to the growing efficacy of political movements among members of middle-ranking cultivating castes, who are increasingly obtaining secure rights to land or outright ownership. During the past three decades, cultivators with land rights have realized that electoral democracy provides them with a measure of political power that their ancestors did not possess, in part because the cultivating castes of India have numbers (and numbers count in elections) and in part because they have leverage over the landless, their only rivals in the villages with sufficient numbers to match them.

The rural landless in India are usually estimated at 26 percent of the total population.[5] Most of the landless come from a few communities that historically have been among the poorest in the world, including the so-called Untouchables or Scheduled Castes (14% of the population), Scheduled Tribes (7% of the population), and minority Muslim community (11% of the population). Neither the Scheduled Castes, Scheduled Tribes, nor Muslims have been able to capture political power at the village and state levels to any significant degree, primarily because they lack the wealth, literacy, education, and other resources that would seem necessary for them to begin to acquire influence in local institutions (schools, courts, village councils, banks, cooperative societies, and so forth). While there have been numerous revolts of landless against the landed, the poorer communities have failed to band together in substantial political movements over any sustained period of time.

There has been a broadening of political and economic power during the past three decades in India, but this has not reached down to the bottom 30 or 40 percent of the population. Pressure has been effectively put on *the* biggest landlords, however, to surrender their control of lands to those who historically have had somewhat insecure land rights. With the introduction of universal adult franchise in 1950, rural politicians have found that they can get elected only if they appeal to farmers who typically have rights to holdings between 5 and 30 acres, or even fewer than 5 acres in some states. All states have now abolished absentee landlordism and set limits on the amounts of cultivable land that can be owned by an individual or family. The most progressive agricultural states have also provided security of tenure to tenants, who were previously entirely dependent on the whims of landlords.

While land reform laws frequently have been violated, either in letter or spirit, time increasingly has favored the smaller and middle landholder. Those landlords who still hold hundreds of acres of land as semifeudal lords invariably have to do so by parceling out their lands among dozens (or even hundreds) of tiny plots, spread across an extensive area. Their principal interest in the land is the collection of rents from sharecroppers and tenants and the increasing value of the land itself. Such old-style landlords are usually not interested in the increased incomes that might come with enhanced productivity. Indeed, they most frequently refuse to consolidate their holdings or improve their land lest the size of their holdings be found out by the land revenue authorities. However, such landlords are finding it increasingly difficult to maintain their holdings without enormous conflict,[6] with pressure emanating from state governments responding in large part to pressure by politicians representing medium and small landholders. Throughout Bihar and Bengal, where such old-style landlords are still common, most of them are sending their sons and daughters to professional schools, with the understanding that dependence on the land alone is unlikely to be enough to sustain them in the next generation or two.

Because of the extreme degrees of scarcity on the subcontinent, India's land problem can be viewed as a double-edged sword. On the one hand, there is not enough land to begin to satisfy the intense land hunger that exists (there would be less than an acre of land per capita were all the land in India—cultivable and noncultivable, rural and urban—distributed equally). On the other hand, unemployment rates are so high that those who do not obtain rights to land usually cannot find meaningful and productive work. Recent studies have indicated that more than 5 million Indians are entering the labor force each year, but that only 500,000 of them are finding jobs in the "organized sector" (i.e., those units of employment in which 10 or more people are employed). The "unorganized sector" is often a euphemism for landlessness in the countryside.

Most of those who have benefited from the spread of political power since Independence are members of middle-ranking caste groups—the Jats, Koiris, Kurmis, and Yadavas are among the most numerous—who historically have been socially inferior to the Brahmins, Rajputs, and other high castes but have always been ranked well above the Untouchables and other Scheduled Castes, the Tribals, and most Muslims. Seldom in positions of leadership, these middle-ranking castes have been classic peasants: trapped in debt by the usurious rates of moneylenders, without secure rights to land, and with little access to the levers of power in society. After the first few elections following Independence, when they divided their votes by giving them to representatives from the aristocratic castes that had always dominated the bureaucracy and educational institutions, these middle caste groups learned to vote as a bloc for their own candidates, and they began to demand more for themselves.

Applying their growing influence in state governments, India's cultivating castes pushed through significant financial measures as well as land reform legislation in the 1960s and 1970s. In Punjab and Haryana, for example, farmers were given large amounts of credit and tax breaks and allowed to import tractors duty-free until local manufacture reached sufficient levels. When the Janata government of Morarji Desai took over from Indira Gandhi's Congress Party in March 1977,[7] they gained concessions from the central government which became so lucrative that they could only be described as price supports. Government money spent on irrigation during the 28 months of Janata rule was more than it was during the entire 11 years of Indira Gandhi's first term in office, or, for that matter, the 17 years of Nehru's tenure as Prime Minister.[8]

Ashoke Krishna Dutt, one of the most sophisticated members of India's Parliament and the scion of an old Calcutta family, summarized the power of India's farmers in December 1978 when he said, "It used to be that we would pass volumes of legislation and nobody in the rural areas would know what any of it meant. Today, these farmers from the villages are coming into the cities and telling us what laws we can pass and what laws we should pass."[9] Dutt was commenting on a rally—the largest ever held in Delhi—that had attracted over a million members of a national rural lobby, the Kisan Sammelan (Farmers' Organization). He might have added that some farmers are not merely voters or lobbyists but number among Dutt's fellow lawmakers. In the 1978 Parliament, over 40 percent of the Members were agriculturalists, a striking contrast with the 13 percent that claimed such status in 1952. In most of India's northern states, the majority of cabinet ministers now come from rural backgrounds.

The farmer who was most outspoken in telling the legislators what laws to pass in 1978 was the founder of the Kisan Sammelan, 79-year-old Chaudhury Charan Singh, who toppled Morarji Desai's government and became India's fifth (and the only non-Brahmin) Prime Minister in July 1979. The son of a poor but landed Jat farmer, Singh speaks in a rural idiom and is in the habit of taking off his shoes and socks, village-style, at public meetings. He has been a champion of farm interests since he first entered political life in the 1930s.

As Chief Minister of India's largest state of Uttar Pradesh in the 1950s and 1960s, Singh was responsible for legislation that effectively abolished absentee landlordism, provided for debt redemption and rural credit facilities, consolidated minuscule landholdings into parcels large enough for profitable irrigation, and gave secure tenancy rights to more farmers than in any other Indian state.[10] As Home Minister and subsequently Finance Minister under Morarji Desai, Singh pushed through a multipronged strategy to set up increases in food production. He also attempted a radical transformation of the Indian economy and redistribution of wealth to India's rural masses.

Charan Singh: Theoretical and Practical Underpinnings

The philosophical basis of Charan Singh's politics is his belief in the possibility of creating a rural "middle class" of peasant landowners and village industrialists. Singh argues that the fundamental problem of our age is excessive bureaucratism and statism.[11] The only way to combat these two evils, he argues, is through the creation of a strong class of peasant proprietors and small local village industries. Singh is opposed to cooperative farming because "it detracts from peoples' initiatives and promotes mechanization at the expense of employment." He is not opposed to service cooperatives (to provide water, credit, and other agricultural inputs); indeed, he is trying to promote them. Peasant proprietorship, in his view, "not only produces more wealth, provides more employment, and removes glaring disparities from the land, but will also prove the most secure base for democracy."[12]

While Singh has attracted the support of many owners of India's larger mechanized farms, for reasons that will be explored later, he himself objects to being identified with them. He constantly quotes figures to show that small farms produce more per unit of land than big farms and is an admirer of the profitable one-hectare farms that are common in Japan. Although he considers forcible land redistribution impractical—one of the reasons Singh has acquired the reputation of being pro-"kulak"—he has worked for consolidation of small holdings into plots large enough to benefit from irrigation, preferably under circumstances where cultivators themselves get secure rights to such land. Singh's ideal is a rural society of millions of prosperous one- and two-hectare farmers, but he also admits the ideal is not possible for present-day India because land productivity is so low.

Singh is concerned about the problem of landlessness among the bottom 30 to 40 percent of India's rural population, but he argues that the vast majority of the landless cannot be given land simply because there is not enough land to give.[13] During the rest of this century, Singh projects a decline in the percentage of Indians owning and working the land, with an increasingly larger portion of the population becoming involved in small-scale and cottage industries. Tracing his aspiration to Mahatma Gandhi, Singh pushed the Janata Party in 1977 to adopt as its economic policy the slogan-like statement that

The Party's policy will be that what can be produced by cottage industry shall not be produced by the small-scale sector and what can be produced by the small-scale sector shall not be open for large-scale industry. The reservation of fields must be clear and, where necessary, statutorially defined. The only exception that may be considered will be for production that is entirely for export.[14]

Fundamental to the support that Charan Singh has generated in the rural areas is the deteriorating position of rural incomes and employment relative to urban progress during the past two decades. The average annual per capita income in the agricultural sector in 1960-61 was Rs. 207 ($25.88) and in 1977-78 was Rs. 633 ($79.13).[15] When adjusted for inflation, however, the real income of the average farmer was only Rs. 201 ($25.13) in 1977-78, actually less than the 1960-61 figure. By way of contrast, average per capita income in the nonagricultural sector was Rs. 577 ($72.13) in 1960-61 and Rs. 2,595 ($324.38) in 1977-78. Allowing for inflation, the per capita real income in the nonagricultural sector rose to Rs. 889 ($111.13) in 1977-78, a 35 percent increase over the 1960-61 figure. The average farmer is thus worse off than he was two decades ago, both absolutely and relatively, while the average worker in the nonagricultural sector has improved his position somewhat.

Such figures obviously have to be interpreted with some caution, since they tend to obscure the skewed distribution of benefits in the agricultural sector. Even though average farm incomes have declined over the past two decades, incomes for the top 10 percent or so of the rural population have risen. This means that the condition of the bottom 30-40 percent of the agricultural sector, which is roughly equivalent to those without land, has been deteriorating to ever more miserable levels.

Farmers who have used the new technology have increased their incomes dramatically over the past 15 years. They do not compare their social and economic positions with those of the poor around them, however. Rather, they now argue there is no reason why they should not have the same standard of living as better-off urbanites. In the words of M. S. Randhawa, former Vice-Chancellor of Punjab Agricultural University:

We should abjure double standards and learn to tolerate prosperity in the rural areas. [Hundreds of thousands] of people in the urban areas own cars, but the mere sight of a farmer on a scooter becomes an eyesore to many people. Our concept of a farmer is a shabbily dressed person wearing a dirty dhoti. Our farmers and the agricultural laborers should live just as well and dress as well as the urban people. Our aim should be to bring rural areas at par with urban areas so far as the standard of living is concerned.[16]

Both the better-off Indian farmer and his poor neighbor are also increasingly discontented because the cost of farm inputs has, on the whole, increased faster than earnings. Agricultural producers often direct their protest at government because it has been primarily through government excise taxes that the cost of inputs has risen. To take the most dramatic examples, the index of aggregated farm input prices in Punjab rose to 213 in 1978 from the base of 100 in 1967. During this time, the price of a 35 hp Massey Ferguson tractor rose from $2,695 to $6,128, the price of the cheapest Indian tractor (Zetor) from $1,712 to $5,422, and the price of a 5 hp electric motor from $123 to $206. In the same time span, the price of fertilizers and agricultural labor almost doubled and the cost of insecticides more than trebled. The price the farmer receives for his product has not kept pace, although wheat-growers have had an advantage over farmers growing rice or other staples such as pulses.[17]

Urbanites generally argue that the farm sector in India is pampered because agriculturalists do not pay income tax, but the farmers contend that they are being overburdened with excise taxes, a wealth tax, and an estate duty. In fact, the power to levy a tax on agricultural incomes is reserved to the states in the Indian federal constitution and only one state (West Bengal) has introduced a bill to levy such a tax. The wealth tax was introduced by Mrs. Gandhi's government in 1970 and affects all lands worth more than Rs. 150,000 ($18,750).[18] As the value of agricultural land has increased dramatically during the past ten years, a number of farmers have been affected by the new tax. Even more important, farmers have been riled by the necessity of getting their land assessed, hiring assessors and lawyers, and spending time in administrative offices and the courts. Similar resentment exists against the estate duty, which affects all land and other property (including agricultural implements) valued at more than Rs. 50,000 ($6,250).

Having glimpsed a vision of the future in which their incomes would be on a par with urban incomes, India's better-off farmers now feel threatened by the possibility that they could become a principal target of taxation by all other segments of society. To mitigate this possibility, rural farm interests in Parliament have demanded "parity," meaning *full equality in rise and fall* between the prices of industrial goods and agricultural produce. As stated before Parliament's Consultative Committee for Agriculture, the farm lobby is proposing that agricultural producers be compensated for increases in their living and production costs to the same extent that industrial workers are compensated by such things as dearness allowances and cost-of-living bonuses. The proposed mechanism for bringing about "parity" is a change in the powers of the Agricultural Prices Commission (APC), which sets government procurement prices for wheat and other crops. According to former Agricultural Minister S. S. Barnala, the Janata government was on the verge of allowing several key industrial indicators to be included in APC formulae for determining agricultural price policies just before the Janata government fell in 1979.[19]

The militancy and potential for violence of India's landowning farmers was indicated in Andhra in December 1978 when rice-growers in Nalgonda district burned cartloads of Phalgun paddy (RP 6-17) in front of food procurement offices, in protest against a government decision not to purchase the variety.[20] Government spokesmen argued they would not purchase Phalgun because the grain has a black tip, which makes it less desirable than varieties with all-white grains. Andhra's rice-growers claimed this was simply a dodge on the part of the state government to diminish its support of farm prices during a period of surplus.

Incidents like the paddy-burning in Nalgonda have become commonplace during the past few years. Even more violent incidents involving landowners—such as one in Tamilnadu in April 1978, when police fired on farmers, killing 7 and injuring 500, when they resorted to widespread acts of sabotage, arson, and vandalism—are increasing in number.[21] In the Tamilnadu case, as in a number of nonviolent demonstrations throughout India in 1978 and 1979, farmers were demanding lower rates for electricity and water in rural areas, reduced prices for fertilizer and other agricultural inputs, postponement of recovery of short- and medium-term loans, and higher procurement prices for foodgrain, sugar cane, and other farm products.

Some of the violence involving landed middle-caste farmers has arisen because the Janata frequently tried to combine middle- and lower-caste groups in political alliances against the old elite castes. The most clearcut case was in Bihar, where Janata Chief Minister Karpoori Thakur based his two-year regime (May 1977 until April 1979) on appeals to the so-called "Backward Classes," a euphemism for middle cultivating caste groups that have been deprived of jobs and educational opportunities. While reservation of jobs and seats in schools may not seem like rural issues to Westerners who are accustomed to thinking of India in terms of peasant stereotypes, for newly wealthy farmers it is now the crux of most political issues.

Thakur (who is himself from a "Backward" caste) reserved 26 percent of government jobs for "Backward Classes," in addition to the 24 percent reserved for Harijans (like "Scheduled Castes," Harijans is another euphemism for Untouchables which means, literally, "children of God"). He also initiated a series of land reform measures designed to put some teeth into land reform laws in Bihar's most feudal districts.[22] While none of the opposition parties in Bihar came out officially against Thakur's moves, upper-caste students in the towns led a state-wide violent agitation in support of the demand that "relief be provided for the weaker sections of the elite castes." As P. C. Gandhi has pointed out, the "advanced castes" felt "their interests [would] be better served if the current turmoil continues and things come to such a pass that the machinery of law and order visibly breaks down beyond retrieve."[23] When Indira Gandhi won in Bihar in 1980, with support from the highest and lowest castes pitted against the middle-ranking castes, most of Bihar's high-caste leaders argued that the students' anti-Thakur agitation had been responsible for the victory.

A second type of conflict in the countryside pits the middle cultivating castes against landless farm laborers and the unemployed, and this type of conflict also seems to be on the increase. Most such conflict comes, as it always has, in local sporadic bursts, when the patience and forbearance of the lowliest castes run out and their anger becomes directed against the newly rich cultivating castes. Attempts have been made—at the local, state, and national levels—to establish permanent organizations of landless laborers that might aggregate some of the discontent that is obviously present, but such attempts have met with, at best, mixed success.

Organizers of the rural landless have encountered two major kinds of difficulties. First, the organizers themselves are invariably from the upper castes and frequently from the urban areas as well. For a variety of complex reasons, the landless have seldom produced their own leadership except for very short-term outbursts. Second, rural labor unions have often been broken by the cultivating castes because rural labor is such a seasonal phenomenon. During the slack periods, a strike or boycott by laborers is ineffective because they are not needed by the cultivating landowners. Planting and harvesting times are so short there are enormous pressures for rural laborers to earn money while they can. If laborers do manage to boycott or strike successfully throughout planting or harvesting, the usual result is a period of scarcity in the village that works primarily to the detriment of the poor. Successful rural boycotts and strikes also encourage the adoption of more labor-displacing mechanization by the cultivating castes and, usually, more and more bloodshed.

There are some observers who argue that a major showdown between the landed and landless of India is imminent.[24] They point to the unquestioned increase in the number of violent incidents where Untouchables have been willing to fight for their rights. They also point to increasing willingness of the rural landless to join in political demonstrations. In this regard, perhaps the most impressive crowd in Delhi in 1979 was an assemblage of landless laborers brought to the city by the pro-Moscow Communist Party of India (CPI) and other leftist parties.[25] Estimates of the crowd numbered from a few hundred thousand to almost half a million. Encouraged by the size of this crowd, India's two major communist parties (the CPI and the much larger and more independent Communist Party of India, Marxist, or CPM) have explored the possibility of forming a "third alternative" between Indira Gandhi's Congress and the Janata Party, with agrarian militancy as a major linchpin. The Janata coalition's disintegration and Mrs. Gandhi's impressive victory in the 1980 election obviously changed the picture, but most observers expect the communist parties to expand their influence throughout this decade.

The Kisan Sammelan

The Kisan Sammelan was conceived by Charan Singh as an organizational alternative to both the communists and Mrs. Gandhi's Congress. Its focus is on nonviolent political activities on behalf of very broad-gauged rural interests. The Sammelan was formed in mid-1977 by Charan Singh and a group of his closest associates as a means to demonstrate the extent of their following. Their feeling at the time was that their rural lobby was not being given its due, by either the Janata Party or the Janata government, primarily because of the common tendency of their opponents to label them representatives of only the richest farmers. Since its formation, the Kisan Sammelan has staged two mammoth rallies in Delhi—on December 23, 1977, and a year later on the same date, ostensibly to celebrate Charan Singh's birthday.

Organizers of the December 1978 rally were quick to point out that it was attended by farmers with large and small holdings, with the vast majority coming from the latter category.[26] The organizers estimated their numbers at 4 million, undoubtedly a gross exaggeration, but the rally unquestionably drew far more than 200,000, the estimate of a spokesman for Indira Gandhi. Police authorities counted more than 95,000 farmers who came to the rally by rail. Independent estimates of those who came by road ranged upward to a million or more, with most officials placing the figure at about 800,000. Many newsmen speculated that the rally might have been the largest ever held in the capital.

A preponderance of the farmers who came to the December 1978 rally were Jats—Charan Singh's own and the largest of the middle agricultural castes—but representatives of other castes and communities were conspicuously present. Both the Sikh Akali Dal and the Muslim Majlis participated. Delegations of retired soldiers, old men and women, children, village wrestlers, and all sorts of other village groups could be seen coming in batches. The largest groups were from Haryana, western Uttar Pradesh, and Punjab, but there were also sizable representations from eastern Uttar Pradesh Bihar, Maharashtra, Gujarat, Rajasthan, and even from far away states like Andhra Pradesh Karnataka, and Assam. For two days, traffic in and around Delhi was completely disrupted, as rows and rows of tractors, trucks, buses, and bullock carts poured into the city. Most of Ring Road, the beltway that surrounds New Delhi, was used as a parking lot on the day before and during the rally. The largest open spaces in the city—the Red Fort grounds, the Ram Lila grounds, and the Ferozeshah Kotla grounds—all became "kisan grounds" for the two days. When the rally started at 11:30 A.M. on December 23, traffic was still stacked up for five kilometers outside the city, most of it farmers trying to reach the center. By early evening, when the rally ended, the traffic jam had been reduced to about a kilometer beyond the city limits.

Singh told the assembled that the Kisan Sammelan would stand for the small, nonmechanized family farmer and landless Harijans. He noted that the organization could not cater only to "kulak" interests because more than 70 percent of India's farmers own less than two hectares each, more than half own less than a hectare, and more than one-third own less than an acre. He advised "noneconomic farmers"—those with too little land to support a family—to shift progressively from agriculture to other professions, pointing out that land is limited and strains on land resources are already considerable. Every farmer, he said, should go back to his village and start a local unit of the Kisan Sammelan in order to put pressure on state legislators and Members of Parliament for greater investments in the rural areas. Only if farmers helped themselves would they be able to convince government that it should do something for them.

Harijans who own land, Singh told the audience, should be allowed to keep it and given secure rights to it. To the extent that redistribution of land was possible, Harijans should also be given land. Singh's emphasis when discussing Harijans, however, was that jobs must be created in the countryside in unprecedented numbers, to be filled by both Harijans and non-Harijans, without land or with uneconomic holdings. He deplored the fact that only 14 percent of the Indian Administrative Service (IAS) comes from village backgrounds. He lamented an expenditure of $20 million planned for a United Nations conference in Delhi in 1980 and suggested the money be spent instead on tubewell or drinking water projects, or on the building of rural roads, schools, and hospitals. "There is no contradiction between the interests of the cities and the interests of the rural areas," Singh told his audience. But then he added, "city people must know that they cannot prosper unless the village-folk prosper."

The bulk of Singh's speech was devoted to the articulation of 20 demands that now form the charter of the Kisan Sammelan, and to a promise to start a "first class" newspaper that will project a rural perspective. The major Indian newspapers, Singh argued, are controlled by capitalists who do not believe in the welfare of the peasantry. This is also the case, he said, for multinational corporations and international agencies like the United Nations.

The 20-point charter started off by expressing "disillusionment" with the Janata Party.[27] "The hopes of the kisans have been shattered to pieces," the charter stated. "They feel bewildered and let down." One of the principal demands called for representation of farmers on all boards and committees affecting rural areas, including the Agricultural Price Commission; commodity boards for tobacco, coffee, rubber, and tea; national fertilizer companies; and banks. Another advocated "parity" between agricultural and industrial prices. A third suggested that import and export policies for agricultural produce be liberalized in order that inputs be obtained at cheaper prices and foreign markets found for Indian farm products. Subsidies for fertilizer, irrigation water, electricity, and other inputs and an exemption from land revenue of holdings below 1.25 hectares were other major demands. Land reform was prominently mentioned, but its meaning was specifically stated so as to emphasize preparation of land records and provision of secure tenancy to cultivators rather than land redistribution.

Farmers and Nonfarmers

As these Kisan Sammelan demands indicate, Charan Singh's aspirations go far beyond representation of rich peasants. He hopes to be identified as the single national figure who articulates the expectations of the peasantry in the mass and stands for the regeneration of rural society as an undifferentiated whole. To be sure, his core support comes from landowners who have outgrown feudal patterns and have adopted modern commercial agricultural practices, but his populist stance against city wealth and his advocacy of industrialization of rural areas has won him support from many nonfarmers in the villages.

Some observers thought for a time that India might have been witnessing the creation of a broad rural front rather than a mere farmers' lobby during the period when Charan Singh was in power as Home Minister (15 months), Finance Minister, and, briefly, Prime Minister. As a *Hindustan Times* editorial (December 25, 1978) put it: "[a Kisan Sammelan] speaking exclusively for the rural areas and articulating a set of comprehensive demands... has every chance of becoming a major political force, bringing on the national scene a distinct political culture which may not always be in tune with modernity." In a similar assessment, noted economic analyst Palraj Mehta described the December 1978 rally as "authentic confirmation of the hold that Mr. Charan Singh has managed to establish on the mass of the peasantry [small, middle, and big], at least in the northern belt of the country."[28] Mehta went on to conclude that "There is a fair possibility that he will extend his presence [among] this vast segment of our people in other parts of the country."

Those who have doubted Charan Singh's ability to realize his dream of a broad front in defense of a consensual rural interest usually focus on the fundamental cleavages of village society as the major stumbling block. They point out that most of the 20 demands in the charter, if enacted, would benefit the richer farmers more than the marginal ones and leave the landless largely untouched.[29] Rural opponents of Charan Singh stress the absence in the charter of any mention of minimum agricultural wages, land redistribution, or other structural change. Urban detractors argue that the charter failed to show any appreciation of the linkages in modern times between investment in industry and investment in agriculture, or the implications of a balanced development of the economy for improving the quality of life of city and village dwellers. Urban economists suggest that subsidization of agriculture to the degree contemplated in the charter would lead to a calamitous retrenchment of public savings. Further insulation of farmers from taxation as recommended in the charter would, in their view, be equally disastrous.[30]

Even many observers who are sympathetic to Charan Singh agree that his most serious political problem is the need for some kind of accommodation between his landed middle-caste supporters and the vast bulk of the landless, who are generally from the lowest (Untouchable) castes. Singh has tried to unite the two by lumping his middle castes among the so-called "Backward Classes" and demanding jobs and education for both the (usually landed) "Backward Classes" and the (usually landless) "Scheduled Castes." Singh argues that nonagricultural jobs in the rural sector should be created at the same rate for marginal and uneconomic landholders and the landless. In practice, however, the landed middle castes have so much power at the local level they are invariably able to garner the most attractive new opportunities for themselves. Moreover, since the landless "Scheduled Castes" usually work for the landed "Backward Classes" as wage laborers, and since the "Backward Classes" are themselves pressing hard for social and economic mobility, the "Backward Classes" have every reason in the world to deny even incremental benefits to "Scheduled Castes."

The 1980 Parliamentary elections seem to indicate that Charan Singh's dream of building a

Table 2

India's Parliamentary Election Results

	1971		1977		1980	
Party	% Vote	No. of Seats	% Vote	No. of Seats	% Vote	No. of Seats
Indira Congress*	43.6	352	34.5	153	42.5	351
Janata	27.6	53	43.2	298	16.3	31
Lok Dal	---(broke from Janata in 1979)---				12.1	41
Communist Party of India, CPI (Pro-Moscow)	4.9	23	2.8	7	2.9	12
Communist Party of India, Marxist (CPM)	5.1	25	4.3	21	4.9	35
Regional Parties	10.5	51	7.0	46	11.5	44
Independents and Local Parties	8.3	14	8.2	14	9.8	11
TOTAL	100.0	518	100.0	539**	100.0	525†

* The Janata totals in 1971 include the totals for the five constituent parties that banded together to form the Janata Party for the 1977 elections. In 1979 the Janata Party divided into two major parties, the Janata and Lok Dal.

**Elections in three constituencies in 1977 had to be postponed because they were located in mountainous areas that are inaccessible during the winter months.

† Elections in 17 constituencies in 1980 had to be postponed because of civil violence in portions of Assam and Northeastern India.

Source: *The Overseas Hindustan Times* (January-February 1980); *India Abroad* (January-February 1980). This article is also dependent, for facts and analyses, on conversations with Professor Myron Weiner of the Massachusetts Institute of Technology; Professor Glynn Wood of American University, Washington, D.C.; and communications from Dr. Bhabani Sen Gupta in New Delhi. For a more detailed analysis of the elections results, see *India at the Polls*, Vol. 2, by Myron Weiner (Washington, D.C.: The American Enterprise Institute for Public Policy Research, forthcoming).

broad rural front against Indira Gandhi and the communists has failed, at least for the present. In those elections, the Indira Congress won only 42.5 percent of the total vote but gained more than two-thirds of the seats in the lower house of Parliament (the Lok Sabha), primarily because the opposition was divided among four feuding national parties, a number of regional parties, and a total of 3,887 independent candidates [31] (see Table 2). In 1977, when all but the pro-Moscow communists were united against Mrs. Gandhi, the Janata Party romped home with a stunning upset victory that included personal defeats for Indira, her son Sanjay, and most of their close associates. The total shift in votes toward the Indira Congress between 1977 and 1980 was only 8 percent of the electorate (from 34.5% in 1977 to 42.5% in 1980), and yet the Indira Congress added almost 200 seats in Parliament!

Charan Singh had a great deal to do with the fall of the Janata government in mid-1979 and the election of Indira Gandhi in January 1980. Although Singh played a major role in the Janata Party's formation—he first broached the idea in Tihar jail in 1976 and the party's electoral symbol (a farmer with a plow) had been that of Singh's Bharatiya Lok Dal—ironically it was Singh who brought down the otherwise stable Janata government in July 1979 by first withdrawing his support from the government and then making an alliance with Indira's Congress. This was universally viewed as an opportunistic betrayal, since Singh had previously been the most vociferous critic of Gandhi and had unsuccessfully tried to jail her (for Emergency "excesses") when he was Janata Home Minister in late 1978. Gandhi outmaneuvered Singh by delivering to him the Prime Ministership in return for his betrayal of the Janata Party, only to desert him and his factionalized party (the Lok Dal) within less than a month's time. Once Indira withdrew her forces from Singh's ramshackle coalition, in August 1979, no new majority was possible and the January 1980 elections became a constitutional necessity.

Since the elections, Charan Singh has been unrepentant about his behavior, even though his rural-based Lok Dal Party has lost the position of dominance it occupied under the Janata regime. Because of his machinations with Mrs. Gandhi, the 79-year-old Singh realized his lifelong ambition to be the Prime Minister of India, if only for 6 months or so. Most observers agree there would have been no other way for him ever to become Prime Minister. The 1980 election results also demonstrated to Janata Party leaders that without him and his landed farmer political supporters Janata cannot win a national election.

Singh's activities during the Janata/Lok Dal period are characteristic of the rather starry-eyed, unsophisticated behavior of many of the middle-ranking rural caste leaders who occupied positions of power in New Delhi from 1977 to 1980. A good many of the new Members of Parliament from peasant stock were thrown off balance by the adjustment to being thrust into office rather suddenly in 1977. As they moved into their posh bungalows in New Delhi—many of them still in ragged *dhotis* and smelling of cow dung—they were often torn between their public defense of rural values and the private temptations of modern urban life. They would condemn urban living, industrialization, and the high life on the one hand, then express their ambivalence by driving around in big cars, trying to get their children into the best professional schools (foreign and domestic), seeking licenses and permits for various forms of government largess, and otherwise building a reputation among their own followers for hypocritical life styles.

Those who successfully resisted the worldly temptations of New Delhi often became, like Charan Singh or Morarji Desai, extremely self-righteous. Combinations of self-righteousness, corruption, and a lack of sophistication about governance led to enormous factionalism within the Janata Party and a general feeling among most leaders that they should seek their own short-run personal gains at the expense of party unity. In this atmosphere, Singh's behavior in seeking the Prime Ministership and, in the process, splitting the Janata Party into several warring factions becomes the more understandable.

The Kisan Sammelan and the Janata Government

It is nevertheless clear that the Kisan Sammelan had an enormous impact on the Janata government. During the Janata regime, India produced record-breaking food surpluses while the government simultaneously increased input subsidies and procurement prices paid to farmers, to the point where foodgrain procurement, intended originally to provide food for ration shops for the poor, became a form of agricultural price support. In order to maintain low ration shop food prices, the government also subsidized the cost of foodgrain to consumers. Some economists have estimated that the total cost to government for these twin subsidies was as much as $1.5 billion per year, half of which was to pay for farm subsidies and the other half for consumer food subsidies.[32]

Janata subsidies, price supports, and other incentives to agricultural producers were the principal factors responsible for India's unprecedented food surplus: production of rice in 1977-78 set a record for the twentieth century in India; wheat production increased by 13 percent in 1977-78, only slightly less in 1978-79. Whereas foodgrain production increased from only 110 to 112 million tons between mid-1971 and mid-1976, it shot up to 130 million tons by mid-1979. Government warehouses had more foodgrain during the Janata period than had ever been the case during any other three-year time span.

A number of economists have argued that Janata's policy of dual subsidization could hardly have been sustained, for it implied that rewards for additional food productivity should go to surplus-producing farmers, at the expense of other segments of society.[33] Indeed, impressive jumps in production, particularly if coupled with stagnant demand and a lack of purchasing power among the masses of the Indian poor, might even have worked to the detriment of India's farming community at some point in the future. This would have been the case, for example, if stockpiles of food had become so large that government and private traders were forced to sell at low prices, thus forcing government support prices to plummet.

The Janata government tried to guard against a collapse in farm prices by increasing food exports in the short run while establishing programs that might in the long run enhance the purchasing power of poorer segments of the population. Rice is still being exported to Mauritius, Vietnam, the Soviet Union, the Gulf countries, and Bangladesh. Wheat is being exported to all of the above countries except Mauritius, plus Pakistan. Negotiations are proceeding on foodgrain exports to Iran and a number of smaller nations in Asia and Africa.[34] Export of other foodstuffs—including meat and vegetables—also expanded rapidly during the Janata period. A task force on agricultural exports which submitted its report to the Commerce Ministry in December 1978 concluded that foreign exchange earnings from foodstuffs and agricultural products which stood at $2.3 billion then could reach as high as $3.9 billion by 1982-1983.[35] The task force recommended that attempts be made to increase export of dairy products by six times its present level by 1982; marine products by two times; and sugar, vegetable oils, and meat products by ten times or more. Traditionally high food exports such as coffee and tea were expected by the task force to experience very modest growth during the next five years.

Critics of the Janata government argue that it should not have exported food when so many people in India continue to be undernourished or malnourished. Indeed, most people are agreed that any talk of food "surpluses" in India must be intended only in a "qualified" sense when the bulk of the population still lacks the kind of nutrition levels found in most major countries.[36] Janata/Lok Dal spokesmen counter by arguing that the purchasing power of the poor can only be enhanced at rates consistent with the absorptive capacity of the bottom-end portions of the economy. While the Janata government invested more in bottom-end poverty programs than any previous Indian government, the reasoning goes, it still found itself left with an exportable surplus of food.

There are definite signs that the Janata government was at least moderately successful in tilting its rural strategy toward small and marginal farmers and some of the landless in backward areas. As a result of unprecedented National Development Plan allocations, irrigation facilities were being added at rates never before contemplated: compared with an average of just over one million hectares of additional irrigation each year from 1970 to 1977, new irrigation facilities added in 1977-78 totaled 2.6 million hectares and in 1978-79 2.8 million hectares. Projections for 1979-80 are for a staggering 3.2 million hectares of new irrigation facilities. The vast bulk of this new irrigation has been in Bihar, Madhya Pradesh, Assam, Orissa, and West Bengal, all of which are among India's poorest agricultural states.[37] Moreover, most of the new irrigation consists of public tubewells and large hand-dug canals, which tend to favor smallholders and those with fragmented holdings rather than the larger landholder with his private tubewell.

The most important programs in building irrigation facilities have been the food-for-work and employment guarantee schemes (EGS) established by the Janata government. The EGS in Maharashtra (which guaranteed work for anyone within 15 days, within 15 kilometers of a village office) generated work equivalent to 150 million person-days in 1978-79, which means year-round employment to over half a million people. It actually affected more than a half-million people because most of the new employment was assumed by the underemployed, who used the EGS to supplement their incomes during periods when no other work was available. Foodgrain allocations to the EGS and other food-for-work programs in 1978-79 were just over one million tons, but the Planning Commission recommended an increase to more than three million tons in 1979-80.[38]

Improved cropping intensity as a consequence of new irrigation was a principal source of new rural employment during the Janata period. The Planning Commission estimated that for every additional one million hectares of irrigation, a half million underemployed were provided with work for three to four months of the year. In addition, programs for introducing new seeds and new technology under the Small Farmers Development Agency to farmers in dryland areas and to farmers who have thus far not used their irrigation potential—were intended to benefit more than two million families per year. Programs for providing year-round employment in cottage industries were expected to add employment for another 1.2 million families each year.

National Sample Survey (NSS) statistics for 1977-78 indicate that there had been (by mid-1979) a deterioration in the rate at which new employment was being created in India's urban areas, but that employment rates for rural women had increased while those for rural men had remained constant.[39] Some opposition politicians attempted to use the 1977-78 statistics to show that Janata's employment-oriented schemes were not working, even though most employment figures for 1977-78 reflect programs in effect when the Janata came to power rather than programs adopted during the new government's first year in office. Figures for 1978-79 indeed revealed a dramatic increase in rural employment during that year. Since Mrs. Gandhi's victory in the 1980 elections, a new Planning Commission has been working on a Five-Year Plan which would provide more balanced investments in urban and rural development, but many of the Janata's more imaginative and popular rural programs are expected to remain on the books. Whether or not they are implemented with the same enthusiasm in the future will depend largely on the new political alignments taking shape after the June 1980 death of Indira Gandhi's son Sanjay, her chief political tactician.[40]

Perhaps the most striking indications of the rural shift that was taking place in the Indian economy under the Janata government came from Planning Commission statistics which indicate that "all of the basic and capital goods industries with input linkages with irrigation and power sectors, and consumer industries which depend on rural purchasing power, have recorded recovery and growth."[41] Among the rural consumer industries that showed the largest increases in production and sales during the two years of Janata rule were cotton yarn, viscose yarn, polyester fiber, cotton cloth, jute manufactures, blades, matches, soap, dry cells, radios, sugar, and cooking oil. Fertilizer use jumped by 25.9 percent during Janata's first year in office and by more than 20 percent during its second year, both records for independent India. Power generation increased throughout India by 11.4 percent per year once the Janata came into office, and consumption of additional power grew much faster in the rural areas than in the urban areas. Indeed, in some of the agricultural states in the northern Gangetic Plain, Janata governments actually took power out of urban areas and gave it to villages.

Evidence of clear benefit to the rural (and urban) poor is provided by statistics on pulse production, which indicate that production of this protein-rich crop (also known as lentils or *dal* in its edible form), a staple of the poor, registered a positive growth of 4 percent per year during the two years of Janata rule. This compares with a *declining* rate of production of 0.6 percent per year during the period 1960-1977. Crucial to these increases in production were drastically enhanced procurement prices and a determined effort to get laboratory knowledge about pulses out into the fields of dryland farmers.[42]

Striking evidence that the Janata government did not entirely neglect the small and marginal farmer was the budget drafted by Charan Singh during his first stint as Finance Minister in April 1979.[43] Singh substantially extended irrigation subsidies to farmers owning four hectares of land or less. At the same time, he significantly reduced interest rates charged by the Agricultural Refinance and Development Corporation for irrigation and land development loans to small farmers. In the new budget, banks were given tax deductions for bad and doubtful debts made by their newly opened rural branches. Budgetary allocations in April 1979 were also generous to Operation Flood II, Verghese Kurien's scheme of dairy cooperatives which is intended to benefit 10 million people in the rural areas, most of whom own cattle but little or no land. Programs for upgrading the cattle population benefited as well from a budgetary provision permitting federal milk cooperatives to deduct from their taxable profits insurance policy premiums paid out by them to insure cattle belonging to members of their village milk societies.

The 1979 budget helped prevent distress sales by poor and marginal farmers, and to a certain extent flattened out seasonal grain prices, by providing for an innovative system of subsidized grain storage facilities run by local communities. Some experiments with rural community cold storage plants—for vegetables, milk, fish, fruit, and other perishable commodities—were also provided for in the budget. Boosts to employment schemes for the landless were the sole intention of a new $35 million allocation for the establishment of weavers' cooperatives and, as well, the abolition of all taxes on state-run Khadi and Village Industries Boards. B. G. Verghese, who had long been an advocate of a shift to an employment-oriented rural development strategy, concluded that all these programs constituted a significant reorientation of the Indian economy. In Verghese's words, Singh's budget "breaks new ground and should be seen as a bold effort to restructure the economy in the right direction."[44]

Charan Singh himself is convinced that some benefits which accrued to India's better-off farmers during the brief time span of Janata rule will also profit smaller farmers and the rural landless over an extended time span. For example, Singh argues that reductions in taxes on agricultural inputs, a fundamental feature of his budget, should encourage the diffusion of new practices among smaller farmers who have a lower risk-taking capacity than the more affluent. Jha has collected statistics to show that increases in overall agricultural output bring down the cost of living for the agricultural worker far more sharply than for his urban counterpart.[45] Conversely, according to these statistics, any incipient shortage of foodgrain hurts the rural poor far more than the urban poor.

The obvious goals of Singh's budget, like the Sixth Five-Year Plan itself, were to curb luxury consumption by taxing it heavily, to avoid putting any hurdles in the way of growth, and to give an edge to cottage and small-scale industries. Almost anything that smacks of a luxury in the urban sector—motor cars and gasoline, air-conditioners, domestic appliances, cosmetics, liquor and candies, radios and stereos, machine-made carpets, steel furniture, and processed foods—all were mercilessly taxed in 1979 at unprecedented rates. Since most new taxes were on consumer products rather than raw materials, the sections of society that were clearly most affected were the middle- and upper-class consumers in the cities and the newly affluent in the rural areas. Surprisingly enough, heavy industry was hardly affected by the new budget, despite Charan Singh's known bias against it.

The 1980 Elections and Beyond

If Indira Gandhi's massive victory in the January 1980 elections was in part a consequence of Charan Singh's ham-handed floor-crossing and factionalism among the opposition parties, it was also related to the attempt by the Janata/Lok Dal governments to shift political and economic power to the rural areas. High agricultural prices, subsidies to food producers, and all the other investments in rural India resulted in a galloping inflation in the cities and neglect of almost everyone in the urban economy. Janata had promised to develop integrated national plans for urban development and energy production, but ended up doing almost nothing for the cities. As industrial production sagged during the Janata years, a number of well-placed economists suggested to Janata leaders that they moderate their rural thrust and begin to make some compromises with the industrial and urban sectors. In the absence of such compromises, it was not surprising that urban businessmen and the middle classes turned back to Indira Gandhi with a vengeance in 1980.

There is also considerable evidence that some key segments of the rural poor turned away from the Janata/Lok Dal parties in the 1980 elections, but not to nearly the same extent as urbanites, rich and poor. Indeed, the Janata and Lok Dal parties elected more Scheduled Caste MPs in the two major states of Bihar and Uttah Pradesh than Indira Gandhi's Congress or any other party! Some of the rural poor clearly moved over to the

CPM in West Bengal, Tripura, and Kerala. This has buoyed the spirits of CPM leaders, who are pursuing a highly innovative political strategy aimed at securing power for both small and marginal farmers and the landless in those three states where the party is now in power.[46] The CPM has been no more successful so far than other political parties in overturning India's traditional power structures at the village level, but most observers would agree that the CPM has the most solid party organization in the countryside. Most would also agree that the CPM has the greatest interest in building support from the poorer segments of the rural population.

Neither the Janata nor the Lok Dal did well in the 1980 elections: Janata secured only 31 of 525 seats, with 16.3 percent of the vote, while the Lok Dal won 41 seats with 12.1 percent (see Table 2). While this establishes the Lok Dal as the second largest party in the Indian Parliament, its position relative to the Indira Congress is so weak that it may have difficulty functioning as an effective opposition. For the time being, at least, the leadership of India's major opposition parties is discredited and dispirited.

The elections have also been damaging to the Kisan Sammelan. The Sammelan still has all the trappings of a permanent national organization—a charter, membership in 18 states, regional offices—but it has been shown to be highly dependent on support generated by the personage and political fortunes of Charan Singh. The Sammelan was unable to build a broad front of medium, small, and marginal farmers, much less to create an effective alliance between these and the rural landless. In the 1980 elections, the support of these groups was split among at least a half-dozen political parties, on the basis of linguistic, caste, and regional factors. With Charan Singh now at the end of his political career, there is no clear successor among the present rural political leaders, and thus Kisan Sammelan too is likely to lose its unique potential and become merely one of many Indian organizations designed to represent rural interests.

This is not to undervalue the role that organizations like the Kisan Sammelan will continue to play in Indian politics. The real importance of the Janata/Lok Dal governments was that they gave a final nudge to a nationwide societal shift that began in the mid-1960s, from a previously half-hearted commitment to institutional reform in the countryside to a more widespread belief in the efficacy of technological improvement. Unlike Nehru, who always argued that land redistribution and other structural changes were necessary before India could contemplate higher productivity in agriculture, Charan Singh and his followers proved to themselves and to many others in India that food productivity can increase significantly without these basic transformations. Whereas Nehru stressed the values of modernization associated with urbanization, and sought to build an industrial technological capability before all else, Charan Singh and his colleagues stress the need for a productive and dynamic rural sector as a base for any future urban or industrial development. In the 1950s and 1960s, Nehru's push for cooperative and collective farming was slowed down considerably by India's farm lobby. In the late 1970s, India was pushed toward an unprecedented emphasis on peasant proprietorship.

Indira Gandhi may now attempt to undo much of the legislation and administrative apparatus that was put in place by the Janata/Lok Dal governments during the past three years. In doing so, however, she will have to be careful not to alienate rural producers to such an extent that they refuse to grow food. This was the case in 1972-73, when Mrs. Gandhi attempted a government monopoly of the wholesale trade in foodgrain, requiring farmers to sell their foodgrain to the government at cheap prices. At that time, food production declined drastically, and it continued to decline during the next few years as a consequence of a series of poor monsoons and further efforts by Mrs. Gandhi to control prices paid to agricultural producers.

If only because of the nature of her own support, Mrs. Gandhi will have to focus her attention on the urban areas. The Indira Congress promised in the 1980 election campaign that it would bring inflation under control, reduce crime, restore morale within the police services, and get industry (especially coal, power, and transport) operating at or near full capacity. To do this, she will have to restore or initiate more solid managerial control of sagging public sector enterprises and discipline errant labor unions. Both efforts will require measures that will invariably be interpreted by her opponents as high-handed.

There are some constraints, however, on the extent to which Mrs. Gandhi can be high-handed this time around. The 1980 elections gave her unquestioned control of the lower house of Parliament but her party still has only 69 of 244

Members of the upper house (Rajya Sabha). In order for her to amend the constitution, she needs the consent of two-thirds of both houses, something she is unlikely to achieve in the near future.

There are other constraints as well. The Janata regime not only repealed many of the constitutional amendments that she had enacted during her 21-month Emergency regime of 1975-1977, it also changed some provisions of the constitution to make it more difficult to declare an Emergency now than it had been in 1975. Moreover, constitutional amendments enacted under the Janata and Lok Dal governments tended to strengthen the positions of both the press and the courts vis-à-vis the Prime Minister's office.

Perhaps the most important constraint on any authoritarian measures that Mrs. Gandhi might contemplate is the control of state governments by opposition parties. Immediately after the 1980 elections, her party controlled only 3 of 22 state governments. While she can be expected to gain control of a number of other states in the months ahead, some state governments should provide political sustenance for the opposition parties for some time to come. As mentioned above, the CPM controls two states and is the leading party in a leftist coalition in a third. The Lok Dal and Janata could conceivably keep Indira's Congress out of office in Bihar, Uttar Pradesh, and Haryana if they find a basis for unity. Regional party governments and opposition coalitions are also possible in Kashmir, Maharashtra, Punjab, and some of the smaller northeastern states.

Some observers would like to depict present political alignments in India in terms of a three-way struggle among the landed, the landless, and the urban populations, with serious class and ideological overtones. In this interpretation, the Janata/Lok Dal parties represent the landed interests in the countryside, the CPM represents the landless, and Indira's Congress represents the urban society and the poor. There is some truth to this characterization, since each of the parties mentioned above derives its greatest strength from the interest group associated with it. But Indian politics is too complex to be explained in such simplistic terms. In fact, all these parties, including the CPM, have had to create aggregative organizations—in which the landed, the landless, and the urban populace are represented—in order to win elections. Despite some recent stirrings, most observers of India's political life are firmly convinced that the Indian poor are at least a decade away from becoming a truly revolutionary force.[47]

In this sense, the importance of the Kisan Sammelan is only in part related to its class base. It is clearly not the only Indian organization that derives support from among the landed. It does differ from some of the others because its founder did make a major attempt, even if it was unsuccessful, to build a broad rural front that included both the landed and the landless. Moreover, the next generation of leadership within Charan Singh's organization is already arguing that the Kisan Sammelan must do in opposition what it failed to do in office—learn to compromise and bargain most effectively with urban-based politicians and representatives of the rural poor.

(October 1980)

NOTES

1. An excellent study of the Indian economy, which emphasizes the increasing poverty of the poorer segments of the rural population as the most important problem confronting India, is Francine R. Frankel, *India's Political Economy, 1947-1977: The Gradual Revolution* (Princeton, N.J.: Princeton University Press, 1978).

2. The most thorough single discussion of agrarian reform in India appears in *The Selected Papers of Wolf Ladejinsky: Agrarian Reform as Unfinished Business*, edited by Louis J. Walinsky (Oxford: Oxford University Press for the World Bank, 1977), pp. 315-567.

3. As Table 1 shows, Indian land distribution is still considerably more concentrated than in South Korea, for example, or Taiwan. In India, 15 percent of those *with land* own 60 percent of the land; even then, there are the large number of people *without* land—about one-fourth of the rural population.

4. A comparison between Punjab and Bihar appears in ibid. See also Marcus Franda, "Extending Punjabi Agriculture Through Youth," *AUFS Reports*, No. 20 (1979); and "Agrarian Reform in North Bihar: Operation Kosi Kranti," ibid., No. 6 (1978).

5. The estimate is derived from the 1971 census figure for "agricultural laborers." Between 1961 and 1971 the category "agricultural laborers," as defined by the census, underwent considerable change. In 1961, anyone who cultivated even a tiny plot of land for some part of the work day as tenants or sharecroppers was classified as a "cultivator" rather than "agricultural laborer." In 1971 the criterion of work was changed to mean "the main activity of the person," or "how he engages himself mostly." Because of this change, the percentage of "agricultural laborers" jumped from 16.71 percent in 1961 to 25.76 percent in 1971, with the latter figure more closely approximating the percentage of landless in the population. Neither is an accurate figure for the percentage of landless in the population; there simply is no all-India statistic that provides such information. In fact, most Indians have some rights to some land, although often to extremely small bits and parcels. Land records are so poorly kept by Indian officialdom as to be almost useless in estimating the percentage of landless or the acreage under the control of landholders.

6. Throughout the past two centuries there has been increasing organization of peasants, particularly among the middle-ranking cultivating castes. The two most prominent Indian leaders associated with peasant movements during the nationalist struggle were Sardar Vallabhbhai Patel and Purushottamdas Tandon. Others, like Choudhary Chottu Ram of Punjab, were important regional figures. Since Independence, a number of leaders have devoted their entire lives to organization of the peasantry, but none has achieved the stature at the national level of Charan Singh in the late 1970s. Among those who come most readily to mind are Professor N. G. Ranga of the Swatantra Party and Harekrishna Konar and Benoy Chowdhury of the Communist Party of India, Marxist.

Most Indian political parties have a peasant wing, which seeks to promote the party program as it relates to peasants. National organizations designed to represent farmers include: the Bharat Krishak Samaj, the National Tonnage Club of Farmers, the Bharatiya Grameen Mahila Sangh, the Farmers' Federation of India, the Young Farmers' Association, the Confederation of Agriculture Relief Association, and the Rastriva Kisan Sangsthan. These organizations are described in a section called "Farmers' Organisations," in *Kurukshetra*, Vol. XXVII, No. 9 (New Delhi: February 1, 1979), pp. 25-27; and *Report of the National Commission on Agriculture, 1976*, Pt. XIV, "Planning Statistics and Administration" (New Delhi: Government of India, 1977), pp. 256-283. See also Braj Mohan, "The Kisan Movement," *Illustrated Weekly of India*, Vol. C, No. 4 (February 3, 1979), pp. 26-29. In two states—Punjab and Gujarat—commercially minded farmers have formed Chambers of Agriculture, modeled after the Chambers of Commerce in the cities.

7. The Janata was essentially a coalition of five smaller political parties that banded together in early 1977 to oppose Indira Gandhi's Emergency regime. None of these smaller parties previously had a national base. The Jana Sangh, the largest of the constituent parties, was associated mainly with militant Hinduism; its electoral strength was greatest in the urban areas and in some small rural pockets of the Hindi-speaking regions. Bharatiya Lok Dal (BLD) and the Organization Congress were heavily influenced by rural landed interests that were anathema to many members of the Socialists, a fourth constituent. The fifth constituent, the Congress for Democracy, was dominated by Jagjivan Ram, leader of India's Untouchable community and by H. N. Bahuguna, a national leader with a fairly strong political base among Muslims.

8. A more complete statement of Janata policies and accomplishments appears in Marcus Franda, *India's Rural Development: An Assessment of Alternatives* (Bloomington, Ind.: Indiana University Press, 1979). See especially pp. 11 ff.

9. Quoted in Marcus Franda, "Farmer Power and Charan Singh," *Asia*, Vol. II, No. 3 (September/October 1979), p. 34.

10. Paul R. Brass, "The Politicization of the Peasantry in a North Indian State," Unpublished paper (September 10, 1977).

11. Singh's philosophy is outlined in Prabhu Chawla, "Charan Singh: Marginal Tactician," *India Today*, Vol. IV, No. 1 (January 1-15, 1979), p. 30. See also Ajit Bhattacharjea, "Choice Before Charan Singh," *Indian Express* (October 15, 1977); and Sitanshu Das, "Charan Singh at Crossroads," *The Tribune* (Chandigarh: June 19, 1978).

12. Ho Kwon Ping, "Singh Takes the First Step to Capitalism," *Far Eastern Economic Review*, Vol. 103, No. 12 (March 23, 1979), p. 75.

13. For a systematic exposition of Charan Singh's position see Charan Singh, *India's Economic Policy: The Gandhian Blueprint* (New Delhi: Vikas Publishing House, 1977). For a critique of this position by an eminent economist and former Minister under Indira Gandhi, see V.K.R.V. Rao, "Charan Singh's Blueprint for India," *The Illustrated Weekly of India*, Vol. XCIX, No. 1 (January 1, 1978), pp. 18-21.

14. The text of the party's economic policy statement appears in the *Indian Express* (August 22, 1977).

15. Figures in this paragraph are from official statistics, as described in D. H. Pai Panandikar, "The Farm Lobby," *Hindustan Times* (January 16, 1979).

16. M. S. Randhawa, "Emergence of Kisan Force," *The Tribune* (January 6, 1979).

17. For a discussion of pricing policy, see my "The Dynamics of Indian Food Policy," in *Political Investments in Food Production*, edited by Barbara Huddleston and Jon McLin (Bloomington, Ind.: Indiana University Press, 1980), pp. 116 ff.

18. Figures in this selection are from M. S. Randhawa, "The Beleagured Farmer," *The Tribune* (October 26, 1978); and *Economic Times* (Delhi: March 10, 1979).

19. Pran Chopra, "Pandora's Box of Farm Prices," *Indian Express* (April 23, 1979).

20. The incident is described in detail in *The Hindu* (Madras: December 26, 1978).

21. See Kasturi Rangan, "Farmers in South India Turn to Militant Unionism," *New York Times* (June 2, 1978).

22. For a description of the attempt, see my "Agrarian Reform in North Bihar: Operation Kosi Kranti," *AUFS Reports*, No. 6 (1978). For an update, see Arun Sinha, "Kosi Without Kranti," *Indian Express* (February 6, 1979).

23. P. C. Gandhi, "Job Reservations in Bihar: Playing for High Stakes," *Times of India* (April 20, 1978).

24. See especially Pran Chopra, "The Sixth Plan's Time Bomb," *The Tribune* (March 24, 1978).

25. The CPI rally is analyzed in detail in *Times of India* (March 21, 1979).

26. Excellent reportage on the Kisan Sammelan rally appears in *The Hindustan Times* (December 24, 1978). For an analysis of its consequences, see Pran Chopra, "The Janata Split and Beyond," *The Tribune* (January 12, 1979).

27. The 20-point charter is reprinted in *The Hindustan Times* (December 24, 1978).

28. Balraj Mehta, "The Charan Singh Phenomenon," *Indian Express* (December 29, 1978).

29. See, for example, the editorial in *The Statesman* (Calcutta: December 28, 1978).

30. Ibid. See also the *Indian Express* (December 25, 1978).

31. Analysis of the 1980 elections is taken from figures and articles published in the *Overseas Hindustan Times* (January-February 1980), and *India Abroad* (January-February 1980). It is also dependent on conversations with Professor Myron Weiner of the Massachusetts Institute of Technology; Professor Glynn Wood of American University, Washington, D.C.; and communications from Dr. Bhabani Sen Gupta in New Delhi. For a more detailed analysis of the election results, see Myron Weiner, *India at the Polls*, Vol. 2 (Washington, D.C.: American Enterprise Institute, forthcoming).

32. D. H. Pai Panandikar, "The Price of Food," *Hindustan Times* (October 3, 1978).

33. Ibid. The Janata government appointed a commission, headed by Vadilal Dagli, editor of the respected journal *Commerce*, to inquire into the subsidies question. The commission was scheduled to release its report in early 1980, but the parliamentary elections of January 1980 intervened.

34. *Hindustan Times* (April 15, 1979).

35. The task force report is detailed in *The Hindu* (December 23, 1978). See also Vinod Mehta, "Export Policy: Accent on Food," *The Tribune* (February 15, 1979).

36. The issue is explored in my "Dynamics of Indian Food Policy," p. 106.

37. For details see R. K. Roy, "New Green Revolution: Some Distributive Aspects of Growth," *Economic Times* (April 11, 1979).

38. Figures are from Planning Commission documents and conversations with people at the Planning Commission.

39. *National Sample Survey, 32nd Round, 1977-1978* (New Delhi: Government of India, 1979), passim.

40. Raj Krishna, "The Budget and the Plan," *Indian Express* (August 7, 1980).

41. Quoted in Raj Krishna, "Performance of Economy: Impressive Gains in 1977 and 1978," *Times of India* (January 10, 1979). The second and third parts of this three-part series (i.e., "Investments Pick up" [January 11, 1979]; and "The Political Shadow" [January 12, 1979]) provide an excellent summary of the state of the Indian economy.

42. For a summary of what is being done, see the special section of *The Hindu* (February 4, 1979).

43. The text of Singh's budget presentation appears in *The Hindu* (March 1, 1979). See also Singh's reply to the budget debate, the text of which appears in *Economic Times* (March 17, 1979).

44. B.G. Verghese, "A Different Kind of Budget," *Voluntary Action*, Vol. XXI, No. 4 (April 1979), pp. 1-3.

45. Prem Shankar Jha, "An Economist's Budget: Clearly Defined Priorities," *Economic Times* (March 2, 1979).

46. The new CPM strategy is detailed in Bhabani Sen Gupta, *CPI-M: Promises, Prospects, Problems* (New Delhi and Stockholm: Young Asia Publications, 1979). See also my "Rural Development, Bengali Marxist Style," *AUFS Reports*, No. 15 (1978).

47. The question is explored in detail in Henry C. Hart, "Explanations," in *Indira Gandhi's India*, edited by Henry C. Hart (Boulder, Colo.: Westview Press, 1976), pp. 278 ff. See also Francine R. Frankel, "Compulsion and Social Change: Is Authoritarianism the Solution to India's Economic Development Problems?" *World Politics*, Vol. XXX, No. 2 (January 1978), pp. 220 ff.

The Farmers' Federation of Thailand

Brewster Grace

In 1974-1976 the Farmers' Federation of Thailand (FFT) blossomed for two years of intense activity.[1] Conceived in 1974, less than a year after an extensive student rebellion helped bring down the oligarchical military regime of Field Marshall Thanom Kittikachorn and General Prapass Charusathien (1971-1973), the FFT rapidly recruited peasant support from much of Thailand, especially in the northern region and central plains. Ultimately, its leadership claimed 1.5 million members. Even allowing for exaggeration by FFT spokesmen, this constitutes a remarkable recruitment and mobilization in a rural society known for its quiescence. Still, the farmers' organization never had the mass following or political clout mustered by peasant movements in certain other developing nations. Moreover, in 1976, with the return of Thailand's military regime, the FFT fell victim to a wave of intense violence directed against all social reformers in the country. Its activities today are minimal or thoroughly disguised. The very existence of the federation—and its subsequent repression—stand nevertheless as unique and frustrating testimony to the possibilities and the limits of populist agrarian organization in that country.

During Thailand's 1973-1976 democratic hiatus from more than three decades of corrupt military oligarchies, the FFT campaigned, recruited, and mobilized on and around issues that went to the heart of the country's political economy and governance. It protested land use policies and practices that it believed to exacerbate fragmentation; it attacked the abuse of land rights by powerful economic interests; and it condemned the continued impoverishment of a majority of Thai peasants because of high rents and low farm commodity prices. The farmers' organization confronted the civilian and military bureaucracies over abuses ranging from neglect and corruption to extortion. The ultimate symbols of social authority, however, were never questioned: the revered institution of the crown and the Buddhist religion.

This last distinction is critical to an understanding of the uniquely Thai conditions in which farmers mobilized for political action. Right-wing groups have attempted to organize farmers, but ultimately they have appeared as noisy, nationalist, anticommunist movements backed by urban and rural business interests and the military. As such, they have lacked strong peasant support. The Thai government has also created farmers' organizations as bases for cooperatives, but these have floundered under the weight of bureaucratic paternalism, inefficiency, and infighting. Finally, on the far left, the Communist Party of Thailand has in some rural areas openly challenged all governing institutions, including the throne, and is fighting a guerrilla war from the mountains and jungles. However, the underlying loyalty of most Thai peasants to crown and religion has served as an effective barrier to widespread communist support. Peasants are also generally aware that the mounting of any radical challenge to the state will bring strong military repression against their village or religion.

The FFT was unique in that it was the first Thai farmers' organization of substantial size and political significance to recruit members on the basis of nonviolent pressure for fundamental change. Many of its leaders were farmers themselves. It avoided confrontation with the revered symbols of national authority, but nevertheless clearly identified social and economic interest groups it considered "enemies" of Thai farmers.

The FFT experience thus illustrates many of the problems and a few of the prospects for "democratic" farmers' organizations in Thailand. To understand its past strength requires a brief overview of Thai society and recent political and economic developments. To understand the FFT's present weakness requires an account of its repression under a restored military government.

Basic Wealth: The Agricultural-Industrial Balance

Thailand's 1976 Gross Domestic Product (GDP) at current prices was an estimated US$16 billion. Agriculture was the largest single contributor, accounting for nearly 27 percent or $4.3 billion, down from 50 percent in 1950.

Historically, rice made up the principal part of this production, and still today land on which it is grown, the product itself, its trade and its milling provide a living for most Thais. Of a population of approximately 42 million, almost one-half grow rice, while perhaps another 10 percent transport, trade, mill, or finance it. While rice is less dominant in Thailand's export economy today, it is still an important source of foreign exchange (see Table 1).

Much of Thailand's rice has been grown in the vast, fertile central plain, flooded annually by extensive monsoon waters from May or June through October (see Table 2). Thai farmers have been able to take advantage of these natural growing conditions to produce the tasty long-grain varieties that command premium prices in markets throughout Asia. Yields have been traditionally low, however, and modern technology has not been able to raise them substantially without the essential precondition of controlled water levels. Drainage systems started in the early twentieth century have only modestly helped to reduce flood damage, and storage for dry season cultivation of high-yielding varieties has not been adequate to increase productivity.

Population growth and good markets throughout this century have stimulated an expansion of settlement and rice production in the mountainous north, and to a lesser extent on the Korat Plateau of the northeast and in the southern peninsula. This expanded cultivation has been most successful in the north, where mountain valleys allow for irrigation systems and the mountains provide for natural water catchment areas.

Table 1

Crop Year	Total Production (tons)	Total exports (tons)[2]	Exports as Percentage of Total Production
1952/53	6,602,000	1,359,000	21
1955/56	8,907,000	1,264,000	14
1960/61	9,475,000	1,576,000	17
1965/66	10,978,000	1,508,000	14
1970/71	13,570,000	1,591,000	12
1971/72	13,744,000	3,113,000	23[4]
1972/73	11,669,000[1]	882,000	8
1973/74	14,920,000	1,200,000[3]	8

1. Below average production due to serious crop failures.
2. Exports during calendar year, e.g., the two harvests during crop year 1952/53, are exported during export year 1953.
3. Government estimate as of July 1974.
4. 1971/72 represented a bumper harvest for Thailand.

Table 2

Irrigated Dry Season Rice Land - 1969 to 1973 (in Hectares)

	North	Northeast	Central	South	Total
1969	2,141	--	35,361	1,670	39,172
1970	2,516	28	44,473	1,369	48,386
1971	2,924	1	63,633	1,176	67,734
1972	2,456	2	82,568	1,164	85,190
1973	7,495	1,719	133,146	1,754	144,114

As a result, water has been more easily controlled and many more farms have been able to plant two crops a year—often using high-yield varieties.

Rice production in the northeast is sadly different. The vagaries of monsoons and the much poorer soil conditions have limited its development and kept most farmers at subsistence levels. In the south, settlement by smallholder rubber producers and Malay farmers has limited the scope for rice production. While the south benefits from two monsoons a year, and some double cropping is possible, its rice surplus is minimal. The rubber, of course, has substantial export value although its contribution to GDP is small.

Maize, sugar, and cassava—all relatively new crops in Thailand—now contribute as much to export earnings as rice. Maize and sugar especially have benefited greatly from expanding world markets. Diversification, particularly to maize, was as much a response to improved market opportunities as a response to the export tax on rice. Known as the rice premium since it was imposed at the end of World War II, the tax has depressed domestic farm prices for rice and promoted the cultivation of other crops.

Timber and opium productions have also grown as a result of world markets but not entirely within the context of agricultural development policy. Opium, grown in the northern mountains by ethnically non-Thai hill tribes, has produced enormous profits—but not so much for the producer as for the international marketers. A hill tribe family producing opium may earn as much as $500 from annual production whereas a Thai rice-farming family may earn only $250. (One pound of pure opium in New York is currently worth $50,000.)

Thai hardwood timbers, most notably teak, have long been major exports. Forest reserves have been seriously depleted, however, and attempts to administer forest conservation policies in the face of strong market demands as well as pressures to clear forest for settlement dim the prospects for serious forest management. Thus, while timber contributed nearly 2 percent to the GDP in 1975, the cost of deforestation, were it calculable, may have been even greater in terms of loss of water catchment areas, erosion, flooding, and the loss of government revenue for licenses which instead are obtained by bribes. Forestry is increasingly an illicit business as reserves decline. Less than 30 percent of Thailand is now forested, a decrease of about 40 percent since 1946.

Thailand's agriculture has provided the base for the industrial sector, which consists, to an important extent, of commodities processing plants. Until the 1930s rice milling was the only significant industry in Thailand. By 1970 there were nearly 50,000 factories registered in Thailand, of which only 26,000 were rice mills. Most of these were small, and fewer than 3 percent employed more than 50 persons.

Although Thailand began to develop other industries in the 1930s, major manufacturing did not begin in earnest until the Promotion of Investment Act of 1954. This allowed for a wide variety of incentives for investment in all kinds of industries. The most notable private, post-World War II capital investors were local, Hong Kong and Taiwanese Chinese, who invested mainly in textiles. Their capital was substantially augmented by Japanese, beginning in the mid-1960s, when domestic labor costs encouraged Japanese textile manufacturers to look for cheaper labor in Southeast Asia.

Such investments are most important in explaining the growing percentage of the GDP in manufacturing. In 1965 manufacturing accounted for 15.5 percent; by 1976 it has grown to 20 percent of the GDP.

Before World War II, expansion of the rice trade allowed for the accumulation of substantial wealth—although much of the profits left the country, mainly repatriated to China. Since the war these funds have been controlled by Thai commercial banks and are the source of most local investment capital in industry.

The most striking long-term investment pattern is seen in the very small percentage of loans commercial banks have made to agriculture. Before 1974, less than 4 percent of the value of all loans went to the farm sector. Farmers instead have had to resort to such "informal" sources as relatives, crop buyers, shops, and moneylenders.

Most of Thailand's recent economic growth has taken place in sectors where relatively few people are employed. Agriculture, which employs at least 70 percent of the population, grew at an average of 4.3 percent between 1971 and 1976, but nonagricultural sectors grew much more rapidly in this period. This differential illustrates a critical problem of income distribution in Thailand today. Urban, white-collar jobs produce far more per capita income than rural employment. Given land shortage, low productivity, persistent poverty, and underemployment in rural areas, income disparities appear to be increasing. So is political instability.

Distribution of Wealth

The extent to which income differentials are changing is hard to determine. Clearly, certain regions and certain kinds of production are becoming more profitable than others. Per capita income for farmers was $103 in 1960 and $139 in 1970. All other workers earned $739 in 1960 and $1,156 in 1970. The actual disparities in income size are relative, of course, to different costs of living and different kinds of income from urban and farm work. But the ratios are increasing: the farmer earned 7.2 times less than the nonfarmer in 1960; he earned 8.3 times less in 1970. The same disparities exist between regions, as indicated in Table 3.

Rice farmers are the largest group of poor Thais today. Historically they have attempted to improve their position by seizing opportunities to cultivate and settle new lands. Today poverty has become an economic and political issue because those opportunities are no longer feasible and cultivation of existing acreage does not produce expected minimum requirements for subsistence.

During the past few years there have been both encouraging and discouraging developments for low income agriculture. Most discouraging has been the constant increase in migration to cities. The Department of Labor indicates that unemployment is now 1.2 million out of a work force of 18 million. Land saturation is perhaps the single most important cause, the squeeze being statistically demonstrated in the decreasing average size of farms. In 1963 the average size farm was 3.2 hectares, in 1970 only 2.8 hectares, and by 1985 an expected 1.8 hectares. Related to this is an apparent increase in tenancy.

There are only a few areas in Thailand where the rural poor have seen signs of improved incomes. In the upper regions of the central plain, areas of double cropping are now developing with farm irrigation facilities from the Bhumiphol and Sirikit dams. In the south, improved techniques for rubber cultivation and recent market recovery have benefited a number of smallholders. Diversification of agriculture, as noted earlier, is progressing at a modest pace, and vegetable, fruit, and maize farmers have all profited from good growing and marketing conditions.

Modern agriculture's promise of higher yields and higher incomes is thus real, given proper program/policy support and an increase of resources entering the rural economy. Efficient government planning, coordination, and stable markets are crucial—but this has not been the rule in Thailand, where middle class bureaucratic and commercial interests have dominated the rural economy.

For the past hundred years the management of marketing has been in the hands of Chinese, long the base of the middle class in rural Thailand. The Chinese in Thailand, unlike those in Malaysia, are relatively integrated. Many have married Thais, most have taken Thai names, and, by and large, they are accepted in rural Thai society. Capable administrators, they have devised a network of buying, transporting, milling, and selling rice that has served the Thai economy extremely well at times of high prices—allowing for rapid expansion of commercial rice. It has also given them a dominant economic position, so that in periods of recession they can usually pass losses back to the farmer in the form of

Table 3

Region	GDP/Capita Increase 1960	1969	Percentage Increase
Central (including Bangkok)	$177	$307	73%
North	$ 79	$117	48%
South	$112	$169	39%
Northeast	$ 57	$ 81	42%

lower buying prices. Whether the incomes of this provincial middle class are ultimately based on exploitative trading is debatable, but its power over the local economy and its contribution to the development of Thai rice production is incontestable. That economic power was dramatically demonstrated in January 1976 when the then Prime Minister Kukrit Pramoj attempted to persuade rice millers and traders to finance rice floor prices through extension of private credit to farmers. Their resistance was enormous and helped precipitate Kukrit's dissolution of Parliament and scheduling of national elections.

The highest income groups in agriculture, if not in rural areas, are owners of vast rice estates, although the extent of holdings still owned by families is hard to calculate. Estimates abound about numerous absentee landlords in the central plains, yet hard data are unavailable. The 1963 Census of Agriculture showed 22 percent of farmland in holdings of at least 10 hectares—three times the size of the average farm. Perhaps the largest landowner is the royal family. Its assets, however, are a closely kept secret and only occasional disclosures by the Land Reform Commission of land gifts by the royal family suggest the extent of their holdings.

Farmers: Friends and Enemies

The FFT and its followers never held the throne responsible for their perceived economic injustices. Indeed, royal patronage a century ago was the mainstay of the Thai social fabric. Feudal nobles were responsible for managing a subsistence peasant economy, thus ministering to peasant needs as they attempted to sustain the court's wealth through control of surplus production. Without legally recognized private landholdings, the peasants were financially dependent on the nobility for access to land. The nobility, in turn, depended on the king for grants of land in exchange for their vassalage. Rooted deeply in this rural tradition is the Thai peasant's inviolable reverence for the throne, as the monarch often protected the peasantry against excessive exploitation by the nobility.

Although the king's power is now widely usurped by commercial, military, and other interest groups, this attitude of reverence has endured. Thus the FFT had no need or wish to challenge the remaining moral power of the throne so long as the king could be supplicated in the name of social justice and the FFT could identify other "enemies."

Buddhism has also remained outside the FFT's explanation for social injustice. Unlike the Catholic Church in much of Latin America, the Buddhist hierarchy has no clearly identifiable class ties, either with the upper classes or the poor. Buddhism as seen in rural villages is socially neutral. The temples and the monasteries serve communal functions; the monks and priests serve spiritual interests: neither condone nor condemn social inequalities. While Buddhist patriarchs may be driven around Bangkok in a Mercedes-Benz, Buddhist monks in villages are poor, dependent on the villagers' alms while preaching basic values of individual fulfillment:[2]

By oneself evil is done.
By oneself one suffers.
By oneself evil is left undone.
By oneself one is purified.

These basic Thai Buddhist teachings, affirming as they do the value of the individual, differ markedly from the tradition of royal patronage, which affirms hierarchy and social difference, though both are perceived to speak to the needs of poor farmers. If Church and royalty are perceived as friends (or as neutrals), who are the peasants' enemies? In the words of Inthra Sribunruang, leader of the northern FFT and himself a farmer, they are "corrupt officials and bloodsucking landlords."

Power in rural Thailand is wielded by those who control the land or its produce and those who represent central, regional, or local political authority.

In the first category are the village headmen who may hold office for life. They are nominally elected by village household representatives. Nominations, however, are made by the village elite, thereby allowing them to control the electoral process. The chief of a group of village headmen (usually eight to ten villages known as a *tambol*) is the highest nonappointed position in rural Thailand. This chief (*kamnan*) is chosen by the village headmen, but his election must be confirmed by the provincial governor. The governor himself is appointed from Bangkok by the Minister of the Interior. The headmen's and kamnan's power embodies the formal authority of all villages and is the critical link to provincial and central authorities. Little transpires between village and bureaucracy without their involvement.

The second level of power is exercised by governors and their district staff who confirm the kamnans. The necessity for confirmation virtually assures supportive village leadership. Moreover, nearly all government projects are administered or supervised by these officials. Agreement and cooperation among kamnan and provincial administration officials are the mechanics of rural power. Finally, these two administrative sources of power are particularly potent when combined with a third power—the absolute power of police weapons. Whether a protection "racket" or mundane commercial transaction, farmers live with a constant threat of violent reprisals should they make efforts to change the prevailing system. The FFT, as we will see, learned this lesson the hard way.

The distinction ordinary Thais make between political and economic enemies on the one hand and traditional patrons on the other helps explain the FFT's weakness internally and vis-à-vis the government. Much of the FFT leadership, particularly that coming from student ranks, became increasingly ideological and hostile to traditional concepts of legitimacy, while the farmers themselves never questioned those basic values. The intensified ideological commitments of the federation leaders (who interpreted Thai politics in terms of class conflict) isolated these activists from their rank and file, and made them and the organization more vulnerable to subsequent repression.

Official corruption and landlord "bloodsucking" have long been associated with patronage in Thailand. Familiar to Inthra and other FFT leaders of peasant origin at the local level, such corruption and exploitation are endemic at the national level. While corruption is commonplace in most societies, it has escalated in Thailand during the past century, largely as a result of the commercialization of Thai rice cultivation, the country's entry into the world rice market, and the development of a modern impersonal bureaucracy to administer the expanding capitalist economy.[3]

Rice exports expanded dramatically from the middle of the nineteenth century to 1940. With this trade came the middleman, his informal "credit" system and commercial rice markets. Commercialization also introduced new landholding patterns not dependent on royal control and patronage. The country changed dramatically within a century from feudalistic Siam to capitalistic Thailand. Thai farmers lost the security of feudal clients and became economic factors in production and capital investment and subjects of bureaucratic control.

It is hard to calculate the exact effects of commercial agriculture on Thai landholding patterns since data are scarce and of varying accuracy. Until the last decade, however, land had been abundant. Between 1850 and 1970, rice cultivation increased 1,000 percent (from one-half million to 5 million hectares), while population grew by "only" 700 percent. Thus Thailand had neither the intense population pressure on the land common to much of Asia nor the large estates dominating the countryside as in Latin America or even in the Philippines. The royal family, as noted, owns vast tracts of land, but some of that property has been distributed periodically to land-hungry peasants as a form of noblesse oblige and an informal (nonlegislated) "land reform."

During the past decade, however, increasing population density on agricultural lands suggests there may be an impending land shortage. Even though the average size of holdings has decreased, the core conflict in the Thai countryside is not the ownership of land per se but the terms of tenancy. That issue has brought many peasant renters into conflict with commercial landlords owning farms of anywhere from 10 to 50 hectares.

A number of surveys have been conducted to establish the degree of tenancy. While none is definitive, all suggest three important phenomena. First, most tenancy occurs in the highly fertile central plain and productive valleys in the north. These were precisely the areas of greatest FFT strength. As many as 50 percent of farmers pay rent. Second, the rent is high—often as much as 50 percent of the harvest. Third, the process of securing land title is so complex that it is a major factor inhibiting change in ownership patterns. Only 40 percent of farmers, according to estimates based on government data, have secure and clear land titles.

Besides its effects on tenancy and rental arrangements, the commercialization of rice created a whole new strata of rural merchants, mainly immigrant Chinese. Partly to offset this economic intrusion and partly to establish new socioeconomic roles, members of the nobility and various court functionaries became civil bureaucrats. In 1932, the king was ousted from governance (but not from the throne) in a military coup d'état. Bureaucrats and commercial rice interests left the king with little power except as a symbol of sovereignty. For FFT and its sympathizers these "usurpers" were the "enemies," with emphasis on one or the other varying from region to region.

In the deltic central plain, where abundant monsoons provide rainfed irrigation to paddies, rice production expanded to become the heart of Thailand's commercial economy. There agricultural producers were set against urban consumers, bureaucratic administrators, and suppliers of services for rice farmers. In the fertile mountains of the north, a similar conflict emerged, exacerbated by population growth that led subsequently to serious fragmentation of holdings. In addition, timber and mining interests in the north were in competition with rice farmers over land and water, abusing both resources through deforestation and water pollution. In fringe areas around the central plains, farmers on small plots felt threatened by large-scale, mechanized production of cash crops on estates of 20 or more hectares.

In the infertile northeast along the Korat Plateau, there was relatively little commercial rice production. Farmers had much lower incomes than in the other regions. Moreover, most of the area's inhabitants maintained a Lao cultural identity. The FFT had little success in the northeast. Nevertheless, the politics of Thai national integrity make the Bangkok government extremely sensitive to any potentially hostile economic and social organization there. Only the Communist Party has succeeded to any degree in organizing the northeastern farmers.

Farmers' Grievances

Historically, Thai rural society has been stable. The FFT attempted to overcome the traditional village and individual apathy to recruit support for its political challenge. The commercialization process and the bureaucracy had provided them three key grievances around which to organize.

First, farmers felt strongly that the government-controlled price for rice had been set too

Table 4

(All figures are US dollars per ton)

	Farm Price[1]	Bangkok Retail Price[1]	Hong Kong Market Price for Thai Rice[1]
1970	$46	--	$191
1971	$36	$145	$172
1972	$47	$149	$195
1973	$74	$158	$426
1974	$90	$225[2]	$626[3]

1. All prices are for the highest grade nonglutinous rice except for the Bangkok retail prices for 1971 and 1972. These are for 5 percent broken which are 5 to 10 percent lower.

2. Average price for January through October.

3. Average price for January through November.

low (see Table 4). The basis for this "antirural" pricing policy has been the subject of acrimonious academic and political debates. Strong arguments have been advanced on both sides of the issue.[4] During the past few decades, observers note, urbanization and agricultural diversification have converted the Thai population from one composed primarily of rice farmers to a nation of roughly 60 percent rice consumers. Therefore, consumer demands for low prices now carry greater political weight than the farmers' desire for higher farmgate prices. Moreover, economic considerations—as well as political ones—weigh against the peasantry. Faced with growing employment needs in the nonagricultural sectors for the past 30 years, the Thai government has had to attract capital into industry. In order to attract capital with a poorly paid industrial labor force (and yet not have excessive labor discontent) the state has used cheap food policies as a means of keeping down the cost of living. Thus farmers have subsidized Thailand's postwar industrialization.

Thailand's modern political economy has created a dilemma for farmers. The nation's poorer peasants are quite aware of their disadvantaged position, but their view of Bangkok is ambivalent. On the one hand, it is an enticing place of glittering lights and off-season (or short-term) employment opportunities. Peasants also realize that it represents another culture whose administration arbitrarily determines their income through the rice market and exploits the rural population for the sake of peace in the cities. Since World War II, when burgeoning urban worker populations increased the demand for rice significantly, the need for low industrial wages to encourage investment (and, hence, the need for cheap food) has become institutionalized in such a way that for a long time farmers could not bring themselves to protest—and the government perceived that it could not allow them to upset the system when they did.

The power of the rice merchants further disadvantages the Thai peasantry. Predominantly Chinese wholesalers, retailers, and exporters have strong family and clan ties whose well administered networks often reach to the merchants' home provinces in China, Singapore, the Philippines, and Indonesia—nations that import Thai rice. Their tight organization gives the merchants a great bargaining advantage, if not actual monopoly, over the native Thai peasantry.

Finally, variable export taxes have been used by the state to adjust price fluctuations in the world rice market to the disadvantage of Thai farmers (see Table 4). When world prices have been low, the farmer has absorbed the loss. When prices rose, the state taxed much of the profits.

Second, throughout 1975 and 1976, the FFT vociferously challenged land rent and land control practices. Tenancy and debt are invariably central to expressions of rural discontent, but the tenancy problem in Thailand is subtle. There is even controversy as to whether it is a problem. There has been substantial research on the subject, a number of conflicting surveys, and numerous hopelessly confusing assertions.

Lawrence Stifel's research tends to reject the conventional wisdom that Thailand's population growth, fragmentation of holdings, and mortgage default—in the broad, historical context of the commercialization of agriculture—has led to increased tenancy. Rather, at least in the central plains, population increases and demand for jobs have been mitigated by urban migration and industrial employment, and the remaining demand for labor opportunities and social equality, it is argued, can be met by technological innovation.[5] Rather, Stifel sees the issue of tenancy as far more complex than simple landlord-tenant conflict. Heterogeneous village patterns reveal numerous variations of landholding patterns, ranging from uncaring absentee landlords to benevolent resident owner-operators leasing their own lands and tenanting others in response to family needs and economic opportunities.

Thai farmers, too, see tenancy within the broader context of rent practices, credit terms, and problems of individual debt, as well as the general political climate. Immediate concerns such as crop failures, access to and prices of fertilizers and machinery, and market prices also deeply affect public perceptions and rural politics. Most important, all these are subject to erratic changes.

Such appears to have been the case in the late 1960s and the first half of the 1970s.[6] Wild price fluctuations, several bad harvests, increasing unemployment, and dramatic inflation in 1973 and 1974 had a profound effect on the small farm economy throughout Thailand, and further weakened tenant farmers' bargaining power in rental agreement negotiations. Almost simultaneously, the student-led revolt against decades of military rule brought unprecedented political liberalization to Thailand in late 1973. The revolt was widely supported by urban Thais (and, at the last minute, by the king). With former Prime Minister Field Marshall Thanom and Deputy Prime Minister General Prapass in exile, the king appointed a caretaker government to oversee the national elections and the writing of a new constitution. The new government was headed by genteel Thammasat professor of jurisprudence, Sanya Dharmasakti. The way appeared to be open for change.

Political activity surged in the months following the revolt, and farmers especially were quick to convert their sense of grievance to political action. In May 1974, several hundred tenant farmers from seven central plains' provinces protested in Bangkok over financial difficulties that were leading them to lose their land. In lamenting the dispossession of their land by creditors, they threatened "to resign as Thai citizens" or "to join the Communist Party in the jungles." (Given Thai political sensitivities, these are the most serious threats possible.) The Sanya government responded by initiating provincial investigating committees consisting primarily of provincial and district officials and a few student activists (see below).

Farmers carried their protests to the provincial capitals as well as to Bangkok in a series of demonstrations between June and November. Chief among their demands was enactment of a state controlled ceiling on rents (a 1950 law set the maximum at 25 percent of the harvest, but the legislation was poorly enforced and rents often exceeded 50 percent). They also demanded government land for landless farmers and title deeds for farmers squatting on government lands. Finally, they demanded a land reform that would limit the size of holdings and transfer land in excess of the ceiling to tenants.

These protests were far from a national mass outpouring of farmers' despair. Yet they were quite significant given deep traditions of peasant apathy and an historic absence of any national or even regional protest organization or network. Most important, the protest occurred despite the marked Thai reluctance to confront the government bureaucracy.

The third grievance, after crop prices and land tenure (including rent), was the government bureaucracy itself. Thai villagers, indeed all Thai, have an ambivalent relationship with bureaucratic authority. The bureaucrat is viewed as a symbol of authority—a representative of the crown—and, at the same time, as an infringer of individual rights. Given good administration, the latter may be tolerable; since bureaucrats more recently were perceived to have served their administrative seniors and themselves at the expense of village and rural interests, dissatisfaction was readily converted into FFT support.

Farmer disdain for the civilian bureaucrats has increased steadily over the years, owing to administrative ineptitude, neglect, corruption, and extortion which result from the bureaucrats' tendency to serve their superiors' needs to gain approval and promotion rather than villagers' needs to gain village support.

Most Thai village development projects involve such substantial capital improvements as roads, dams, and irrigation canals. These require village labor which, in turn, demands village organization. In building a dam, for example, an official of the Ministry of Interior or Royal Irrigation Department needs to turn to the government-sponsored local farmers' organization or co-operative. More often than not, however, this is only a paper organization; peasants don't acknowledge their co-op commitments. The government official is then left with little choice but to command villagers to work as "volunteers." As an official he usually has the clout to do so, but the price is more resentment, for the villagers then want wages that were not budgeted into the project's cost.

The literature on rural development in Thailand is replete with examples of bureaucratic incompetence and peasant disaffection.[7] An early

1974 incident, to cite but one, involved nearly 2,000 acres of land in northern Chiengmai province. Farmers who cultivated it claimed the land had been given to them by the provincial government several years earlier. They were suddenly informed that a corporation representing a Thai-Israeli joint venture had secured title to the land and that they were henceforth dispossessed.

The farmers did not leave quietly, however. Their protests to the Chiengmai provincial government brought about the establishment of an official investigative committee. By that time, however, the Provincial Land Office had already claimed there had been no "irregularities" in the corporation securing the titles. The investigating committee, in turn, could find no basis for reversing the commercial transaction because the land had been "allocated" to the villagers over the past ten years and "investigating allocations which were made so long ago is a problem as the documents involved are now hard to trace."[8] Ultimately, the farmers gave up; they were offered jobs on the new corporate farm and on less desirable adjacent government-owned lands.

Chiengmai province was a likely location for such an incident, as land pressures are most severe in the north. While this appeared to have been a spontaneous protest, it is not surprising that the FFT later found many of its adherents in the region.

Extortion is more difficult to investigate and prove. Still, Thais firmly believe that government officials, particularly the military and police in the northeast, are often engaged in protection rackets. As noted, the northeast region has more than average difficulty in relations with the central government. Due to its remoteness, its poverty, and disaffection, civilian development programs and communist suppression campaigns are closely interwoven. "Winning hearts and minds" often takes brutal forms. Especially during 1973-1976, the civilian government was forced to investigate police killings of peasants accused of being communists. In one instance, investigators claimed that 21 villagers had been murdered for refusing to pay local police protection money.

The rest of Thailand does not match the corruption or other "wild west" qualities of the northeast, although poor farmers elsewhere often feel, and indeed are, vulnerable to violence—including that committed by the government. The FFT's meteor-like success in 1975 and, equally important, its decline in 1976 offers much evidence of this.

Organizing the Farmers' Federation

A number of economic and political interests emerged and converged to produce the FFT in 1974. Perhaps the most important of these were the many small farmers' groups in the central plains and the north which had formed and maintained their existence despite the rigorous constraints of the military governments. The groups were organized around very specific local issues—more often than not to protest forced resettlement because of irrigation or dam projects, or to oppose loss of land due to large-scale cash crop projects. These local protest groups subsequently produced the organizing leadership of the FFT.

Student activists helped the local peasant leadership to come forward and engage in national politics. When the political liberalization began in 1973 a small number of aggressive students went into the Thai countryside to mobilize incipient peasant organizations. Their peasant contacts taught the students their basic lessons in rural economics and rural politics. The following farmer's story told to student reporters was typical:

I have been farming since I was 20 years of age. I initially owned only two acres of paddy field and had to rent another two acres. When I was 22 years old, I tried to buy myself more. I applied for the loan from the cooperative to buy about one and a half acres. The cooperative could not loan me enough. So I was forced to borrow the rest from a moneylender with high interest rates. In my farming, I had some good crops; but I suffered failures in other years and lost both my original paddy fields and the newly purchased land. Finally I went bankrupt and had to quit farming when I was 32.[9]

For the first demonstration in Bangkok in June 1974, university students helped charter buses for farmers, particularly those from the central plains provinces. Student activists' headquarters at Thammasat University became centers for protest planning and coordination. Saman Luang, a public commons outside Thammasat University's main gates and less than a mile from the National Assembly, became the protest site and the farmers' encampment.

These campaigns were not successful in mobilizing mass national farmers' "marches on Bangkok" as promised by students. Not more than 1,000 came for the first rally in June and it is doubtful that more than twice that number came in November. The demonstrations were unique in Thai politics, however, and achieved their objectives by gaining publicity and sympathetic support from the press and left-of-center politicians. Moreover, by the end of 1974 the Farmers' Federation of Thailand was organized as a broad coalition of leaders from the geographically disparate local organizations (farmers from the central plains and the north still dominated the membership, however).

The protest rallies also produced critical political concessions from the newly installed Sanya government. These, in turn, enabled the federation to expand rural support. Sanya promised a land reform law (albeit quite vague) by 1977 as well as several temporary measures to protect farmers against usurious interest rates. The government also promised to end evictions of existing squatters on government lands.

Most important, Sanya's Cabinet rushed through the National Assembly a rent control law similar to the rarely enforced Act of 1950 which put a ceiling on rent at 25 percent of the harvest. The main difference was that the 1974 law was bolstered by grievance investigation committees known as "*kaw saw saw*" consisting of provincial officials, landlords, and tenant farmers. Satisfied, the FFT was induced to return to the countryside where federation leaders began to draw aggrieved farmers' attention to the new rent law and the government's other promises. Pressure was shifted from the central government in Bangkok to rural authorities and local vested interests. Thus, FFT leader Inthra Sribunruang noted:

Concerning farm rent, we gathered in Bangkok demanding that the government promulgate a Farm Rent Control Act, which they did last year. However, the villagers generally don't understand the essence of the Act and how they can benefit from it. I myself travel extensively trying to explain the essence of the Act to Federation members and all villagers in every district in Chiengmai and Lamphun provinces.[10]

"To explain the essence of the Act" as stated by Inthra, was to encourage and mobilize tenants to confront landlords. When FFT organizers and farmers took their grievances and evidence of violations to the kaw saw saw, landlords were enraged. Provincial officials—often friends of landlords—were placed in embarrassing positions. Confrontation was often direct. Individual farmers demanded adherence to the rent control legislation, and rapidly growing new chapters of the FFT simultaneously posed threats to vested village economic interests in land ownership and control. The FFT's tactics were nonviolent—it sought to use existing legislation and institutions such as the kaw saw saw to achieve its objectives. Its leaders traveled relentlessly, talking to farmers and consulting with them. Within 12 to 18 months, the FFT had become a significant rural force, even though it may never have achieved the 1.5 million membership claimed by its president in 1976.

The FFT never formally supported any of the political parties that proliferated during 1974, although it was courted by several. Left-of-center party leaders often appeared at the Bangkok rallies to address the assembled farmers. Others, politicians more right-of-center, criticized the growing politicization of the farmers' movement. Eventually, the FFT came to be associated principally with the Socialist Party and especially with Boonsanong Punyodyana, a U.S.-educated sociologist at Thammasat University who ran for parliament on the Socialist Party ticket in Chiang Rai in the 1975 elections.

There was a certain "chicken or egg" quality to the pattern of FFT politics after 1975. With its somewhat reluctant but possibly inevitable involvement in politics came increasing confrontation with rural authorities and landed interests. By the end of that year, the FFT had become the object of organized and violent opposition. Had it been able to remain purely a farmers' movement, the hostility of its opposition might not have spread to national political activities nor involved such intensive ideological debate.

First, Farmers' Federation members and organizers were sometimes barred from public or village meetings, while landlords and village headmen refused to publicize the contents of the 1974 Land Rent Act or refused to establish or participate in the kaw saw saw. At the same time, local police and military began threatening to arrest as a "communist" anyone voting for or supporting the FFT. Indeed, many farmers and other rural activists were arrested and detained. The most prominent instance was the arrest of nine farmers and students in the north accused of destroying private property in an FFT action against a mine. (The mine's waste water downstream was polluting irrigation.) The arrests may have been legally justified; but the subsequent political explosion engulfed even the prime minister—who ultimately intervened and pardoned all nine.

Local landlord resistance to the Sanya concessions and subsequent sabotage of his reforms and harassment of the FFT escalated quickly and violently. The FFT responded in two ways. First, its language and propaganda became vitriolic and infused with class jargon. Inthra Sribunruang, one of FFT's more vocal spokesmen, wrote in the "Thai Farmer":

Where there is such oppression, there is also struggle. Who can withstand such inhuman conditions...?

Victory will come only from fighting. Corrupt officials and bloodsucking landlords have become frightened by the strength of the FFT so now they are fighting like dying men....

Our struggle is just, therefore people will support us. Let us make friendly gestures towards others. When we gain more strength, we will have victory in the end. Our struggle will take time and be protracted, but we will have the final victory.[11]

Second, the FFT decided to bring their campaign from the village back into national politics. A May Day protest was held in Bangkok in 1975 to dramatize rural conditions and to focus attention on the promised land reform, which had yet to be worked out in detail. The protest, however, was badly organized and the government called FFT's action illegal. So with badly bruised feelings, the protesters left Bangkok on May 2 to return to the villages.

Sanya, meanwhile, had been replaced by an elected government headed by Kukrit Pramoj in April 1975. The new cabinet reflected powerful old commercial and military constituencies. A brilliant politician, keen intellectual, and prolific political commentator, Kukrit built a coalition of diverse constituencies. Since his party held only a very small percentage of seats in parliament, compromises were a necessity. Whereas Sanya did not have to deal with a reluctant, once-bitten twice-shy military and right-wing establishment, Kukrit was embroiled in their reawakened concern for threatened self-interests. Sanya's concessions to the farmers were replaced by political compromises reflecting Kukrit's need for support from the vested interest groups. Thus the FFT's May Day protest fell on ears deafened by growing economic and military concern for "national security" and by Kukrit's political insecurity.

Regardless of their strategic retreat, national politics would not ignore the FFT. Emerging right-wing political groups gathered rural support from village chiefs and tambol kamnan, from the police and military, and from Bangkok business and commercial interests who attributed the economic downturn to the general political instability. National political groups such as the "Nawaphol" found ready support from rural goon squads to harass farmer and student activists associated with the FFT. The prestigious Thammasat University became a bombing target of right-wing student groups. Much of the FFT's leadership, it appears, was assassinated systematically. By the end of 1975, at least 26 FFT organizers had been killed, among them the outspoken Inthra Sribunruang, who was shot on July 31. (Inthra had been warned by the governor of Chiengmai province that Nawaphol's "killing units" were hunting FFT leaders in the North.[12])

There is not much that Prime Minister Kukrit or his delicately balanced coalition government could have done to stop these killings, and much of Thailand's law enforcement was passively sympathetic to the harassment. The powerful military and commercial interests elected to parliament in April 1975 had no incentive to do anything about it. Moreover, Kukrit was weak and needed an expanded base. Policies conciliatory to FFT and its left-of-center supporters were impossible. Instead, Kukrit sought support in rural areas by making compromises with the growing opposition to progressive reformists. The FFT was forgotten in the shuffle, left as prey for the right.

While ignoring the FFT, Kukrit did not ignore the needs of rural Thailand. He attempted three rather imaginative programs. First, he allocated $250 million as direct grants to tambols; second, he pressed into parliament for a national, rural election law which would require a formal democratic election of all village leaders and kamnan; and third, he sought to increase the farmgate price of rice by establishing a floor price which all middlemen buyers, including government agencies, had to pay farmers.

Theoretically, the first program rather neatly by-passed the bureaucratic hassle of village project funding. Tambol councils were told "here is money for you. Design your own village project and it will not be subject to local officials' arbitrary control." Second, as part and parcel of this effort, Kukrit's desire for democratic election of village leaders represented genuine concern for village opinion as to how the money should be spent. District and local officials and life-long village chiefs were excluded. Third, the Prime Minister's establishment of a farmgate floor price for rice was an attempt to achieve more equity between farmer and consumer.

The FFT, however, saw no value in Kukrit's programs. The tambol funds and rural elections were viewed, respectively, as pork barrel or welfare projects and as constituency building for Kukrit's Social Action Party—an unimpressive parliamentary minority. The farmgate price policy included a number of difficult and complex compromises with consumer interests—including food stamps—which left the FFT, at most, unenthusiastic.

Thus, Kukrit gave the federation little to be excited about while its membership was ravaged in the countryside. Boonsanong Punydyana, like Sribunruang was also killed. The FFT in general became a victim of the passion directed against any movement that threatened Thai social order. Village headmen were organized by the Nawaphol to protest and ultimately to subvert the democratic election process. Tambol funds evaporated with inflation and corruption. Middlemen in the rice market combined with the general urban population in a strong protest that defeated Kukrit's attempt to pay farmers a higher price for their rice. Indeed, the price issue ultimately brought down the government on a vote of no confidence. In the ensuing election in April 1976, Kukrit Pramoj lost. Rightists and business interests had coalesced against him for his failure to restore law and order to the countryside.

Thailand's brief experiment with democratic government ended with a military coup on October 6, 1976, a violent reaction by traditional Thais to disorder and disorganization. The average Thai felt the situation had been generally out of hand and welcomed the law and order promised by the new military rulers.

Conclusions

The FFT shattered the historic Thai assumption that farmers will always coexist peacefully, passively, and productively with officials and rural economic elites. It also challenged the assumption that political activism in rural Thailand was inevitably communist agitation. With these assumptions gone, decades of development activity and administration are open to question. Can large bureaucracies, their highly trained administrators, and their physical infrastructure address the basic needs of Thai farmers—land reform, rent control, and higher crop prices—without the kind of participation by farmers that the FFT provided?

The first problem is convincing Thai leadership that FFT-type organizing efforts need to be encouraged and protected from reprisals; once national leaders are convinced, the next problem is how to implement peasant participation.

The National Extension Improvement Project of the Ministry of Agriculture and Agricultural Cooperatives, for example, attempts to relate all extension work to Tambol Farmers' Associations, the government-sponsored farm organizations. The project places so much emphasis on farm extension administration, however, that it looks to the peasant like another formal institution.[13] There is nothing in the project that allows truly independent farmers' organizations to develop projects and participate in rural reforms—as attempted in the kaw saw saw.

The last problem is perhaps the key to handling the first two. What are the surviving FFT organizers and sympathizers to do? Some are now convinced that democratic reforms through populist farmers' groups will not work and that the correct course is revolutionary change. There are others whose perspective is more pragmatic: learn from the experience and look for opportunities for action within the present political framework. Most of the former have joined underground political movements; those with the second outlook are back in their usual occupations. The armed revolutionary movement in Thailand is small and very weak, although it continues to put pressure on military governments for agrarian reform by providing an alternative for Thais disenchanted with their economic and social situation. It is not unreasonable to speculate that Thailand will continue to find that these disenchanted groups cannot be served by more bureaucratic institutions and administration. Similarly, it is possible that many former FFT members may join and influence the government-organized Farmers' Associations. If this is the case, these two alternative views could both serve the same purpose—to convince the government to promote and protect genuinely independent farmers' organizations working for agrarian reform.

NOTES

1, Much of this paper is based on the unpublished analyses of Bantorn Ondam, "The Farmers' Federation of Thailand: The Right to Organize." Professor Bantorn, a rural sociologist at Bangkok's Thammasat University, has methodically reviewed all FFT literature and accounts of repression of FFT organizers. Without his documentation this paper could not have been written.

2. For one of the most succinct statements of religious values and rural development in Thailand see: George Rosen, *Peasant Society in a Changing Economy* (Urbana, Ill.: University of Illinois Press, 1975), pp. 129-168.

3. The two major studies in English on the Thai bureaucracy are by Fred Riggs, *Bureaucracy and Political Development* (Princeton, N.J.: Princeton University Press, 1963), and William Siffin, *The Thai Bureaucracy, Institutional Change and Development* (Honolulu, Hawaii: East-West Center Press, 1966.

4. The "rice premium"—an export tax on rice—has been subject to heated scholarly debate. The most succinct analysis is by Ammar Siamwalla, "A History of Rice Policies in Thailand," *Food Research Institute Studies*, Vol. XIV, No. 3 (Palo Alto, Cal.: Stanford University Press, 1975).

5. See Laurence Stifel, "Patterns of Land Ownership in Central Thailand During the Twentieth Century," unpublished manuscript.

6. See Chaigong Chuchart, "Proposal of Land Reform and Development in Thailand," Department of Land Development, Bangkok, 1975.

7. See Herbert Rubin, *The Dynamics of Development in Rural Thailand* (DeKalb, Ill.: Northern Illinois University Press, Special Report No. 8, 1974).

8. *Bangkok Post*, February 16, 1974, p. 1.

9. Bantorn Ondam, op. cit., p. 4.

10. Bantorn, Ondam, op. cit., p. 19.

11. *Bangkok Post*, July 18, 1975, p. 3.

12. Ibid.

13. See United States Agency for International Development, *Thailand—Agricultural Extension Outreach* (Washington, D.C., 1977).

Rural Mobilization for Modernization in South Korea

Albert Ravenholt

Dramatic physical transformations in South Korea's countryside, now brusquely hauling ancient villages into the late twentieth century, are the culmination of a rural revolution let loose by Japan's surrender nearly 35 years ago on August 15, 1945. That first popular upheaval was thwarted, as the United States Army of Occupation instead opted for stability and utilized much of the centralized Japanese colonial government machinery, often with the same police and other Korean personnel. However, it did prompt the beginnings of land reform, a change desperately needed in this crowded peninsula. When an independent Republic of Korea government, elected under United Nations auspices in 1948, did adopt a comprehensive Land Reform Law in April 1950, implementation was crippled by the terrible trauma of the Korean War that engulfed the country on June 25, 1950, and by clumsy administration. Korea lacked political leaders of stature to guide reform and Americans of competence comparable to those who transformed Japanese agriculture during the Allied occupation. Nor did the U.S. Congress prove perceptive with its assistance. Its program in Korea contrasted sharply with policy in Taiwan, where the United States had chartered and funded the Chinese-American Joint Commission on Rural Reconstruction that largely designed and monitored the model land reform, acting independently of domestic pressure groups. The failure to match land reform in Korea with creation of effective farmers' cooperatives to provide credit, extension, marketing, and other means of mustering rural self-help severely retarded agricultural progress. Korean peasants thus struggled to survive until the late President Park Chung-hee in 1961 established his "command system" of Five Year Economic Plans, initiating rapid industrialization for export and enhancing employment. Yet, it was only 10 years later with the beginnings of this urban prosperity that South Korean farmers could garner the full benefits of owning their land, as the *Saemaul Undong* (New Village Spirit Movement) utilized new technology to modernize production and rural life.

Land hunger among the peasants of East Asia has been endemic for centuries, and Korea is no exception. The critical equation is the relationship of population to arable fields, and modern statistics reveal part of the conundrum. The 23 percent of Korea's total land area that is cultivated—just under 5.5 million acres—must support approximately 38 million inhabitants, or about 7 people per acre. Since North Korea is about one-fourth larger in area and holds only approximately one-half of this population its dilemma is less severe. Japan and Taiwan have the same man/land ratio but a friendlier climate. India, by contrast, holds about two persons per cultivated acre and resource-rich Burma has two acres of arable land per capita. The United States farms about 1.5 acres per citizen and utilizes about twice as large a range area for livestock, both to feed its population and to sustain the world's largest agricultural exports.

It was the accumulation of such population pressure in Korea and other agrarian societies in Asia that generated land tenure problems and abuse of the peasantry. When no additional fields could be reclaimed from forest or fen in the valleys, peasants seeking land crowded up the slopes. But reclaimed upland fields usually lacked irrigation and in drought years left the tiller's family hungry. Given their precarious economic status, many small farmers became indebted to moneylenders, lost their property, and became tenants paying rent to till fields of those more fortunate or powerful. An ancient adage in Asia says where the land is most fertile, the concentration of people will be greatest and the land rents highest. As rents escalated, so did prices of land, leaving even less opportunity for young men who did not inherit property to accumulate the means needed for buying their own farms.

Just as the Black Death liberated the surviving poor in fourteenth- and fifteenth-century England by enabling them to command better pay for work, so plague and other humanly costly disasters and wars periodically eased pressures upon Asian peasants. But the respite was brief; ironically the genius of traditional agriculture that enabled Asia to accumulate roughly one-half of the earth's population also fostered a pernicious tenancy system. Successful land reforms of classical eras, when emperors at the center sought to curb the power of dukes in the hinterland by redistributing their lands to tillers, were set aside by forces the peasant could not escape. The emperor still had to contend with the avarice of the court, its concubines and tax collectors and the cost of the armies. Moreover, peasant sons could be conscripted for *corvée*, constructing the palaces, moats, walls, and canals demanded by the rulers. In Korea, as in China and Japan, the gentry known as the *yangban* profited by the peasant's subordination. Perhaps more than elsewhere in East Asia, the gentry in Korea were parasitic; the rents they collected supported a life of often scholarly leisure yet almost never produced innovators in agricultural production, as did the gentry in eighteenth- and nineteenth-century England and some regions of continental Europe.

The first meaningful figures on tenancy follow Korea's formal incorporation into the Japanese empire in 1910. For the Japanese, imbued with the fervor of expansion, Korea was "an economic missionary enterprise." By 1921 the Bank of Chosen (Korea) proudly reported that in their first decade Japanese administrators had increased the arable land, through land reclamation, by 76 percent to nearly 11 million acres, allowing an average of about 4 acres per family for the then predominantly agrarian population of 16 million. The Japanese introduced their system of "model farms" to demonstrate agricultural innovation, started seedling stations, and established an agricultural and forestry school. Production of most major crops, from oats, millet, barley, and soya beans to silk cocoons, pears, and apples, increased substantially. In that first decade rice production increased 214 percent with improved fertilizer supply, irrigation, and better varieties.[1] The value of rice exports increased nearly ten times. Although Korea's rice producers also received somewhat higher prices, this betrayed Japan's primary objective, which was to utilize Korea to feed the burgeoning population in the home islands, especially after the "rice riots" following World War I, and make it a land for profitable Japanese investment.

Established with Japanese government backing even before the annexation of Korea, the Oriental Development Company acquired huge tracts of land. So did private Japanese individuals and smaller firms. In the 1920s a Korean educator found in the most favored rice-growing region of a southeastern county that some 8,000 Japanese owned two-thirds of the total value of assessed property, while 120,000 Koreans owned only 32 percent.[2] From 1914 to 1937 the percentage of owner-farmers decreased from 22 to 18. 9, part-owners who also rented some land decreased from 41.1 to 28.4 percent, and full tenants increased from 35.1 to 52.7 percent. The terms of tenancy were even more critical. Rents throughout Korea averaged 50 to 60 percent of the crop. Tenants paid all farming expenses, including water charges, seed, fertilizer, taxes, and fees for inspection, plus occasional "gifts" to landlords. Some tenants were left with less than one-quarter and sometimes only one-sixth of the crop in return for their labor.[3] As in Taiwan, Japan, and some other regions of Asia, the tenants also were victimized by the pernicious practice of oral rent contracts requiring annual renewal, which allowed the landlord or his agent to extort additional contributions.

Koreans today recall the era of Japan's colonial administration as one of privation and humiliation. For many it was. Partly because the Japanese administrators thought it was too cold for their own people to settle, after their conquest of Manchuria in 1931 they pushed Koreans to this largely frontier region of Northeastern China. Other Koreans sought escape from drudgery by migrating to work in less desirable jobs in Japan, where they became a depressed and despised class. Japanese pricing policy in Korea forced most Koreans to eat barley, millet, sweet potatoes, and other "rough foods" as staples, while fine white rice went to the local elite or was shipped to Japan.[4] While this diet was healthy and, along with their beloved pepper-hot

Table 1

Farming Households in Korea According to Status, 1914-1937

Category	1914 1000's	1914 Percentage	1924 1000's	1924 Percentage	1937 1000's	1937 Percentage
Landlords	47	1.8	102	3.8	–	–
Owner-farmers	570	22.0	526	19.5	543	18.9
Part-owners / Part-tenants	1,066	41.1	934	34.5	814	28.4
Full Tenants	911	35.1	1,142	42.2	1,511	52.7
Total	2,594	100.0	2,704	100.0	2,868	100.0

Source: Data for 1914 and 1924 from *Agrarian Problem and Peasant Movement*, Vol. 4, p. 42, published by International Agrarian Institute, Moscow, 1937; data for 1937 from Johnson, A. A., "Farming Households, Holdings, Ownership, and Tenant Status in Chosen," in *Civil Affairs Handbook, Korea*, p. 25.

Table 2

An Estimate of Land Ownership Distribution in Korea in the Pre-World War Period

Classification	Number	Area Owned (in 1,000 acres)	Percentage of Total Farming Population	Percentage of Land Owned to Total Land
Landlords	83,000	7,198	2.7	63.9
Owner-farmers	502,320	2,888	16.3	25.7
Owner-tenants	729,320	1,174	23.6	10.4
Tenants	1,583,435	0	51.3	0.0
Laborers	116,020	0	3.8	0.0
Kadenmin *	71,187	0	2.3	0.0
Total	3,085,282	11,260	100.0	100.0

* *Kadenmin* are transient agricultural families which squat on government- or privately owned forests or wasteland, clear the land by burning, cultivate the land until its fertility is exhausted, and then move on to a new area.

Source: Grajdanzev, *Modern Korea*, pp. 111-12.

kim chi (pickled vegetables), probably helped sustain the Korean's physical vitality, it was resented. After the start of the Sino-Japanese War in 1937 and especially after Pearl Harbor, Koreans were conscripted into Japanese labor battalions, police ranks, and support units of the Imperial Army and sometimes vented their own resentment in brutality against other occupied peoples of Southeast Asia. All Korean society was traumatized by Japanese efforts to erase the Korean culture and language.[5]

August 15, 1945

As news spread through Asia that the Emperor of Japan had accepted the Allied terms for surrender, it triggered euphoric celebrations such as these peoples had probably never experienced before or since. Pent-up hatred of the Japanese was replaced by joyous and often unrealistic expectations. In a belated compromise agreement between Moscow and Washington, D.C., Russian troops who had been at war with Japan for less than two weeks occupied Korea down to the 38th parallel; American troops of the 14th Corps, U.S. 10th Army, only established themselves in Seoul by the 9th of September 1945. In theory, the division of the Korean peninsula was to be temporary until Japanese military and civilians were repatriated and a trusteeship administration was established in keeping with

President Franklin Roosevelt's ideas. From the time Russian and American military forces entered the country, however, they almost inevitably began creating two different societies. Meanwhile Koreans abroad were hurrying home by any form of transport they could manage; about 3.5 million returned to South Korea and increased its population to more than 19 million. Over a million had come from Japan and an equal number came from Manchuria and North Korea, including refugees who fled for fear of the communists, plus those from China and even America. Bruce Cumings has ably pictured the events in his article on liberation in the Kyongsang provinces of Southeastern Korea, 1945 to 1950:

> *In the atmosphere of utter release coming on the heels of the Japanese demise, local governing committees, peasant unions, labor unions, and youth organizations spread throughout the Korean peninsula and inaugurated years of virtually unprecedented mass political participation. In South Korea this movement was under the aegis of the Committee for the Preparation of Korean Independence* (Kōn'guk chunbi wiwōnhoe), *established on August 15, and its successor, the Korean People's Republic* (Chosŏn immin konghwa'guk) *established on September 6.*[6]

The most enduring were the People's Committees that initially seized authority in most communities throughout South Korea and often continued to challenge U.S. military administrators when they arrived, especially when they reinstated Korean policemen who had served the Japanese.

In this environment it will be evident why land reform had already begun in Korea before the Japanese emperor announced surrender. As Japanese—and there were over 700,000 of them in Korea as administrators, businessmen, etc.—saw their "rising sun" fast setting, they hurried to secrete their wealth. The best available estimate indicates that as of 1942 Japanese owned 781,788 acres of farmland in South Korea—more than 15 percent of the cultivated area and usually the best quality land. As defeat approached Japanese tried to sell their property, or at least place it in the hands of a Korean, who might be an employee or mistress. Private Japanese property, whether houses or land, not so transferred often was seized by Koreans shortly after liberation. Within less than a month of its arrival the U.S. Army, on October 5, 1945, hinted at what was to come with issuance of Military Government Ordinance No. 9. The statute aimed to help Korean tenant farmers with four provisions:

1. Farm rents were fixed at a maximum of one-third of the total value of all crops produced on the farm.

2. Where rents were less than this maximum, they were not to be increased.

3. All written lease contracts were to be registered at the local land office within 60 days; where lease contracts had been oral, they were required to be put in writing and similarly registered.

4. Rents no longer were to be paid in kind, but rather in cash at the legal, government-set price of those agricultural commodities, or at the reasonable existing market price of the product.

In the chaotic circumstances of early postwar Korea the U.S. Military Government (USAMGIK) lacked both the precise data and administrative machinery to supervise enforcement. In February 1946, USAMGIK drafted a "homestead act" that would have given Korean tenants title to the land they tilled. They were to pay 3.75 times the value of annual production over 15 years by turning in annually 25 percent of their main crop. Leftists challenged gradual conversion, insisting upon outright confiscation, which the Americans were unwilling to do with Korean property and decided instead to leave the decision to a Korean government to be established. However, Korean landlords, many of whom had been identified with the Japanese, saw the shape of things to come and a significant number made private sales to their tenants, sometimes on the most onerous terms they could arrange.

USAMGIK, on February 21, 1946, issued Ordinance No. 52 and thereby created the New Korea Company to acquire and manage all property of the Japanese Oriental Development Company and other Japanese rural lands in South Korea—one-seventh of agricultural land—on which the new tenancy protection statutes were enforced. Ordinances Nos. 173 and 174 (March 22, 1948) passed this responsibility to the newly created National Land Administration, which promptly initiated a "land-to-the-tiller" program. The maximum family holding was set at about 2 hectares, or 4.9 acres.[7] Land in excess of the ceiling was to be sold to experienced farm laborers, including refugees from North Korea. Land prices were fixed at three times average annual production; because they calculated values in the controlled price of grain, however, the real price was 180 percent of annual production. At that time the average price was equal to the value of 131.58 bushels of rice per acre. Tenants purchasing the land had 15 years to pay and those payments were allowed to be part of the quota grain sales to the government. They were forbidden to resell the land before the mortgage was paid or for 10 years, whichever was longer. Crop failure was the basis for a moratorium in payment and land could not be lost to other debtors. The National Land Administration continued to provide low interest loans for production credit and family emergencies. Within months the number of owner-farmers increased by 126 percent over the World War II period and this well designed and managed partial land reform sent a surge of interest and hope through South Korean rural villages. Meanwhile, after carrying out their first land redistribution in North Korea, where per capita farmland was much more abundant, the Communists had reshuffled occupancy of land. Some farmers had been deprived of their holdings while parcels of land were given to those who proved politically more faithful, although rhetoric indicated this was but a stage toward collectivization.

When North Korea thwarted the United Nations from holding supervised elections north of the 38th parallel, the United Nations Commission proceeded with elections in the south on May 10, 1948. This led to formal establishment of the Republic of Korea in the south on July 12, 1948, and five weeks later the formal ending of U.S. Army occupation. Chronic guerrilla activity had plagued many areas, sparked in part by refusal of the occupation authorities to accept the legitimacy of popular, often leftist-led, and largely peasant-based People's Committees. These Committees, claiming governmental powers, struggled with the more rightist elements often associated with collaboration. The influx of so many newcomers in a time of extensive unemployment also contributed to unrest. The USAMGIK land redistribution had nevertheless proven immensely popular in the countryside. Candidates in the May 1948 elections found it expedient to favor redistribution of large private landholdings. The "Constitution of the Democratic Republic of Korea," adopted after the election, included a provision that "farmland shall be distributed to self-tilling farmers. The method of distribution, the extent of possession, and the nature of restrictions of ownership shall be determined by law."

The administration of President Syngman Rhee and the National Assembly wrestled with this mandate as both tenant and landlord groups fought for their interests. Law No. 31, passed by the National Assembly and promulgated on June 21, 1949, was amended on February 10, 1950, by Presidential Decree No. 275 and on March 25, 1950, by Decree No. 294, both designed to clarify administration and execution. Briefly, it provided for government purchase of all land not tilled by owners and any land owned by farmers in excess of about two hectares. Locally elected land committees were to fix compensation to the previous owners at 150 percent of average annual production of the fields, with a reduced compensation rate for larger holdings. Compensation was made with negotiable land bonds redeemable in five annual installments. Present tenants had the first option to buy this land at the same price and terms of payment up to the legal limit of about two hectares. After that the option passed to farmers with smaller areas capable of cultivating more, families of those who died for Korea (as in resistance to the Japanese), farm laborers, and those repatriated from abroad. Violation of the law or proven deceit could cost a landlord compensation for his land and a tenant his rights to cultivation, plus severe fines. Landlords could donate property only to charitable and educational or other public institutions; they could not circumvent terms of the land reform law by giving parcels of land to relatives and others.

War in Korea

It is one of the ironies of history that North Korea's attack upon South Korea on June 25, 1950, appears to have been prompted at least in part by these very land reform measures. As the popular, largely peasant-based political forces won concessions from the established order, even imperfectly, the government's legitimacy increased and it succeeded in pacifying former guerrilla strongholds in the south. Communist leaders in the north apparently decided they no longer could look to internal upheaval in the south and must thus resort to outright military conquest. The North Korean forces usually introduced their version of land reform into the areas of the south they occupied; its provisions left the landlords 12.25 acres each if they were actually farming, the same allotment allowed in the north where land was more abundant and the population sparse. Soldiers also had to be fed, however, housed (often in the best rooms), and kept warm with fuel. The *corvée* labor demanded of all peasant families included in time even women and older men, who were denied rations and policed in gangs. As Sidney Klein has observed:

Aside from the infringement on their personal freedom, farmers in occupied South Korea felt that in effect one member of the farm family was working directly for the State indefinitely without compensation, and they objected to it very strongly.[8]

The ruthless conscription of rural folk for labor, combined with land redistribution favoring the families judged most politically allied with the communists, left an indelible memory in the minds of South Korean farmers, as did their discovery that in the north land ownership certificates were withdrawn and reissued only to those who had cooperated enthusiastically. Once they discovered all the exactions demanded, many in the south concluded the communists essentially wanted them to be tenants of the state.

Yet, the brief communist occupation of much of South Korea in summer and fall 1950, and again after the Chinese Red Army entered late that year, made it simpler for President Rhee in October 1950 to proclaim resumption of land reform. Harsh treatment by the communists who confiscated land and appreciation that the South Korean government was paying them for land lost now made landlords more amenable. A pattern for land redistribution had already been established by the very efficient sale of former Japanese properties initiated by the USAMGIK on March 28, 1948, through the National Land Administration. By September 15, 1948—within less than 7 months—505,072 families of the 582,853 families farming plots formerly belonging to the Japanese had received 197,417 hectares. The balance of 271,211 hectares either was in the process of being distributed, was wasteland, or was in disputed ownership. Amid the battles of the Korean War, the new government proceeded to distribute land formerly owned by large Korean landholders through the adjudication of elected land committees on which tenants and landowners had equal representation. Within 2 years 314,270 hectares were distributed to 1,001,694 families; an additional 234,864 families appear to have been unable to pay for their allotted 102,330 hectares. While the U.S. Army-initiated distribution of former Japanese properties had benefited 29 percent of the then farm households in South Korea, now another 44 percent of farmers had benefited. In addition, near the end of World War II much land had been sold privately to tenants by landlords.

By 1954 statistics compiled by the Republic of Korea government and calculations by Americans who had worked on these efforts indicated roughly one-half of all South Korean farmers owned the land they cultivated. Nearly 40 percent of farm families owned part of their land and leased the balance under protected terms. Only 10 percent of the rural population were full tenants or laborers. Within less than a decade there had been a fundamental social and economic transformation in rural South Korea. Nearly 2.5 million acres had been redistributed to 700,000 peasant cultivators.

Table 3

Estimate of Changes in Land Ownership in South Korea, 1945-1954

Percentage of Total Farm Population

Classification	1945	1954
Owner-farmer	13.8	50.4
Half-owner	16.3	31.2
Half-tenant	18.4	8.1
Full tenant	48.8	7.2
Other	2.7	3.1
Total	100.0	100.0

Source: Estimate for 1945 from U.S. Department of State, "Land Reform in Korea," in Mitchell, *Final Report and History Of The New Korea Company*, p. 262.

South Korea's experience, however, differed significantly from the dramatically successful rural development programs in post-World War II Japan and Taiwan. Dr. Chiang Mon-lin, the former chancellor of Peking National University (1919-1945), was chairman of the Chinese and American Joint Commission on Rural Reconstruction (JCRR) that transformed agriculture and village life in Taiwan. More than two decades ago during a field trip visiting farms in Taiwan he summarized to this writer what had been learned. "We must walk on two legs," Dr. Chiang emphasized. "Land reform alone is not enough. Unless the farmer also has adequate and reasonable credit, reliable irrigation, effective extension of new technology and marketing of his produce at fair prices, owning his land can become a mirage. Most of these services can best be

Table 4
Rural Households by Land Area
South Korea 1974

Percentage of Total*

Hectares	Rural Households	Arable land
under 0.5	29.7	10.0
0.5-1.0	35.7	28.3
1.0-1.5	19.2	25.2
1.5-2.0	8.6	16.0
over 2.00	6.8	20.5
Total	100.0	100.0

*From data in "Population and Food in Korea," by Sung Hoon Kim and Dong Min Kim (1975), p. 29. All other basic data are from annual "Reports on the Results of Farm Household Economic Survey."

Source: H.C. Kriesel, "Food and Nutrition Policies in Agricultural Development," in *Papers on Agricultural Development Policy Seminar* (Seoul: Ministry of Agriculture and Fisheries, 1977), p. 24.

supplied through cooperative farmers' associations that are soundly managed."

In South Korea, with its harsher winters and shorter growing season, such rural facilities are even more critical. After land reform a representative rural family possessed two acres to cultivate: roughly 2.5 million rural households shared 2,240,000 hectares of arable land. Fortunate was the farmer with at least one-half of his land irrigated to grow rice and the balance available for raising barley, sweet potatoes, vegetables, and other crops primarily for household consumption. But such small farms, especially in regions where the weather allows growing only one rice crop annually plus possibly a winter crop of barley, demand intensive inputs of fertilizer and much else to produce a living for the cultivator and his family. It was lack of access to credit and means to such inputs that inhibited roughly a quarter of a million Korean tenant families from immediately availing themselves of the government's land reform in 1950; later some of these families were able to acquire land. Despite the massive American and other foreign assistance after the Korean War,[9] however, there was no appropriate and effective organization, like the JCRR on Taiwan, for delivering these necessities to the countryside. Thus South Korea's agriculture stagnated throughout the rest of the 1950s.

The Korean War (1950-1953) was nevertheless a turning point in the modernization of South Korea's agriculture. On the one hand, the loss of human life was terrible.[10] Physically, much of Korea was devastated, as only modern warfare can cripple production and accomplish destruction of bridges, houses, orchards, or livestock eaten by the soldiers intent on survival. Yet this cruel crucible of war had a constructive by-product. In the minds of the predominantly rural populace it introduced a technological revolution. Boys from the farm who formerly had plowed with an ox discovered that mechanical skills paid high wages, especially when they could repair or drive a 10-ton U.S. Army truck. The KATUSA, or Korean Augmentation to the U.S. Army, itself was a revolution. By bringing Koreans into American army units—the U.S. Second Division now at the 38th parallel at last report still included 8,000 KATUSA or Korean soldiers—encouraged not only weapons skills but also knowledge of the entire new world of electronic communication, use of heavy construction equipment, and organization. As a Korean officer corps was tested and emerged from the war, members went on to master modern management methods at military schools mainly in the United States. Given the Korean propensity for leaders to emerge from small, close-knit, elite groups with shared loyalty and commitment, they provided the drive and direction that is so evident in this almost breathlessly modernizing nation.

A New Order

The stumbling administration of President Rhee, which lacked a coherent agricultural development policy, was replaced in the military coup of 1961 that brought President Park Chung-hee to power. The new president and his associates planned for economic development with a ruthless drive comparable to a general staff preparing for a D-Day invasion, using organizational techniques often learned at U.S. Army military management schools. Options for any mass-based pressure groups of farmers, students, and others to compel government action in their interest were largely set aside. Instead, technocrats came to the fore, joining their foreign-acquired expertise to the now military-dominated political and administrative government structure. As South Korea embarked upon its successive Five Year Plans, agriculture was initially deliberately subordinated to extraordinarily rapid industrialization with major emphasis upon supplying export markets. It was a calculated choice of priorities prompted by the assessment that South Korea's natural physical resources were meager, while her human population offered a most energetic, naturally skilled, low-cost, and productive labor force. As happened in several other countries of Asia, war had shattered the old, almost feudal, hierarchical social structure, encouraging those who were mechanically and physically adept, stimulating geographic mobility and greatly expanding the domestic market economy. In effect, the government chose a centrally planned "guided capitalism" that borrowed techniques for mobilizing production and marketing from the powerful international Japanese trading companies.

During the decade from 1961 to 1971 South Korea averaged an annual economic growth rate of better than 10 percent, with then only modest inflation. Manufactured exports grew about 55 percent annually. This rapidly expanding urban economy created "spill-over" benefits for farmers. As purchasing power increased among factory and office workers they had more money to spend on staple grains, vegetables, fish, fruit, and meat. At the same time, young people now went off to seek employment in the cities, and while wages were often low they still enhanced cash available to their family in the countryside. Population shift reduced the rural proportion of the population from about 60 percent to 42 percent and like Japan at that time, South Korea was becoming "a land of old farmers." The Office of Rural Development (ORD) was removed from the Ministry of Agriculture and Fisheries and made directly answerable to the president, with both the agriculture and home ministries providing support to some 8,000 extension workers. Through this mechanism provincial governors and other officials were enlisted in helping organize the "winter short courses" that might enroll farmers three days to a week. Each short course emphasized a specific crop and in some years up to 200,000 farmers attended—where possible they met at experiment stations.

Early in the 1960s the semiautonomous government agency, the National Agricultural Cooperative Federation (NACF), extended low-interest loans to farmers in order to ease the burden of private indebtedness, only to discover that many families used the funds for nonproductive purposes. (It can be argued that had there instead been professionally managed and genuinely farmer-controlled cooperatives, as in Japan and Taiwan, neighbors in the countryside might have policed each other.) The NACF was also utilized to supply chemical fertilizer and farmers were made mutually responsible for each other's payment. During the first three Five Year Plans (1961-1976) chemical fertilizer use per hectare increased from 72 kilograms to 260 and the rate of pesticide application increased 18 times. Assistance went to the building of reservoirs and village irrigation associations. During the first Five Year Plan agricultural income grew by 5.3 percent annually, but during the second Five Year Plan it dropped to 2.5 percent yearly. The living standard of farm families was falling behind.

Rural Mobilization

Mild stagnation in rural incomes and a worrisome rate of rural to urban migration convinced President Park and his ruling associates that something more effective had to be done in that countryside. After the military coup on March 16, 1961, then Major General Park, as chairman of the Revolutionary Committee on National Reconstruction, reminded fellow Koreans that he had been a farm boy once and pledged to improve rural life. Likewise both the officers and enlisted men of the military, upon which both political power and the defense of South Korea depended critically, came chiefly from the rural areas. During the first five years following Japan's surrender Korean peasants had shown forcefully that they no longer would suffer peacefully. Possession of their land—only about 7 percent were now left as full tenants and they enjoyed protection of their leases—had fortified rural political confidence. Discouraged farmers in Korea as elsewhere have a way of simply not growing much. Wages and salaries in the cities had approximately doubled in the 1951-1961 decade and most families spent most of this for food. During that period South Korea's imports of grain had multiplied more than 6 times to about 3.2 million tons in 1972. The government still sought to restrict rice consumption, requiring restaurants, for example, to observe two "riceless" days per week, substituting barley, millet, wheat products, or tubers, all of which are cheaper staples and less prized than rice. Thus in South Korea, as in the rest of East Asia, the goal of self-sufficiency in rice production has both a psychological and a strategic importance unknown in the Western World.

As the South Korean leaders and technocrats approached the quandary of how to enable rural villages to "catch up" with the rest of the economy—and produce more rice at the same time—they could capitalize upon several national assets. Prosperity in the industrial-export sector already was generating national income that provided capital and encouraged confidence among international lenders. Koreans had a national commitment to education matched in very few countries. They worked with a will that even the Japanese and the Chinese in Taiwan, Hong Kong, and Singapore still do not quite equal—in 1978 the average South Korean wage earner worked a 56-hour week. Irrigation had been extended to 86 percent of the 1,230,000 hectares of paddy fields, just under half the total hectarage (2,240,000) cultivated at the close of the decade. Between 1969 and 1971 the first really new rice varieties since the Japanese had introduced theirs half a century earlier became available in South Korea. Developed in cooperation with the International Rice Research Institute (IRRI) in the Philippines, these varieties incorporated the *japonica* characteristics adapted to northeastern Asia and preferred by consumers there. There still remained the problem, though, of how a Korean farm family (now cultivating an average of .94 hectares, or 2.32 acres) could be enabled to make a decent living.

The solution chosen was the creation of the *Saemaul Undong*, or New Village Spirit Movement." In effect the government wanted to inhibit possible organization for rural discontent, as had happened earlier, by mobilizing all 34,815 villages for rapid increases in agricultural production and modernization of living throughout the countryside. Elements of this program were accomplished by the government directly, such as extension of electrification to the hinterland and construction of modern highways linking farmers with their markets and suppliers and the larger national economy. But the basic need was to enthuse farmers and their families with the conviction that they themselves could make the difference. It all began modestly after 1970 when each village was given 350 bags of cement, produced by South Korea's now booming cement industry, on condition that they muster the local labor for community improvement. Where a village proved they could use this well, additional assistance was offered and other villages were informed they would be similarly assisted when they demonstrated comparable initiative. Next was the election in these villages of representatives to be sent for training at Saemaul Undong Institute at Suwon, some 30 miles south of Seoul. Here the training emphasized the sharing among these elected leaders of knowledge about how their villages had successfully mobilized its citizens for popularly shared goals. Villages competed with their neighbors to move in graduated categories from "underdeveloped" to "developing" and on to "developed," when streets and bridges were being rebuilt and most of the houses improved and topped with tiled roofs, partly financed with government loans. When this program was extended to the cities late in the 1970s, there was considerable grumbling by urbanites who felt it an imposed and onerous burden even though it remade sections of their neighborhoods.

Similarly intensive methods were utilized to induce villagers to follow the examples of model farmers and Four-H Club leaders who pioneered in using new seeds for intensive cultivation. By 1976 about 44 percent of all paddy fields were planted to the new varieties of rice, and South Korea became self-sufficient in production of rice with what proved that year to be the highest per crop average yield in the world, 5.3 metric tons per hectare. Two and even three rice crops annually allow southern Taiwan and some growers in the Philippines and elsewhere in Southeast Asia to produce a larger total harvest in a year. Overall, however, Korea ranks fourth in the world, right behind Spain, Japan, and Taiwan and far ahead of the Philippines, in average rice yield per hectare[11] —achieved in part because the long days in summer that South Korea enjoys between 33° and 38° north latitude allow for greater photosynthesis during that season.

As labor shortages began to appear, there was new emphasis upon mechanization of land preparation, threshing, and even transplanting and harvesting rice. Four-H Clubs often led the way in one- or two-day national rat control campaigns. Rice drying and storage also were modernized and milling was improved to produce a remarkable 72 percent return of milled rice from dried rough paddy rice.

Specialization in farming became a major movement, as demand escalated for poultry, fish, shellfish, meat, fruits, and nuts like the chestnut, which Koreans savor. Windswept Cheju Island in the Yellow Sea expanded growing of tangerines and other citrus to a total of 130,000 tons per year. Hog raising also became a commercial enterprise for many farmers who talked eagerly about the merits of crossbreeding with imported boars. Increasingly, Korean farmers have been raising dairy and beef cattle, chiefly starting with foundation stock from New Zealand and Australia. Both sericulture and fish culture have become major commercial ventures. South Korea's landscape is dotted with the plastic greenhouses such as are common in Taiwan and Japan, growing out-of-season vegetables and flowers. The ancient practice of *kadenmin* (slash and burn) in government- or privately owned forests by transient families was brought strictly under control. For the first time in many centuries the countryside was mobilized not only to plant trees on the mountain slopes but also to protect them from animals and fire. Among the heroes of the *Saemaul Undong* are farmers who have reclaimed wasteland, sometimes with little help in government loans and extension services, to grow mushrooms, apples, or oriental pears.

Prices received for their produce are fundamental to the attitudes and behavior of farmers. Like so many governments in Southeast Asia and elsewhere, that of South Korea had followed a policy of keeping down the cost to the consumer of rice and other staples. Beginning with the U.S. Army occupation, growers were required to sell to the government an annual quota at controlled prices and could sell the balance on a free market. Consumers at that time also relied for part of their diet upon rations. In the early- and mid-1960s the system of direct government purchases was gradually replaced with control of pricing; the government attempted to increase consumption of both domestic and imported barley—including the 25 percent restaurants then were required to mix with rice—by subsidizing the price to consumers. Only in 1968 was the policy reversed and prices paid farmers for rice that year increased 17 percent, to be followed by a 22 percent increase in 1969, a 35 percent increase in 1970, a 25 percent increase in 1971, and annual graduated increases since then. During the first three Five Year Plans ending in 1976 the price of rice to the farmer increased more than 12 times, calculated in *won* (the post-devaluation exchange rate in June 1980 was 591 *won* to US$1). Throughout the decade of the 1970s the government attempted to maintain farm prices for rice at close to the equivalent of US$.70 per kilogram (2.2 pounds) for polished rice, which in most years was about twice the world average for this grade. By contrast the price paid to farmers in Taiwan for rice was about 40 percent lower than in South Korea and that paid to farmers in Japan about twice as high. Most of these increases in rice prices have been passed on to the South Korean consumer and have become a factor in inflation. On the other hand, price incentive certainly contributed to South Korea's record 1976 harvest, which averaged 5.3 metric tons of milled rice per hectare.

Just as improved prices, together with the inputs mustered as part of the *Saemaul Undong*,

Table 5

Main Food Production for Selected Years, 1961-1976*

(in thousands of metric tons)

Crop	1961	1966	1971	1976
Rice	3,463	3,919	3,998	5,215
Barley & Wheat	1,389	1,845	1,714	1,847
Pulses	190	195	263	354
Potatoes	383	972	707	671
Misc. cereals	96	107	110	119
Fruits	150	331	404	615
Vegetables	1,235	1,716	2,908	3,217
Special Crops **	47	60	89	85

* During the time periods the land use pattern has changed due to the pressure of urbanization and industrialization. Some paddies and croplands were used as industrial sites, while river and tidal basins were reclaimed for agricultural use. Nonetheless, overall hectares of the land under cultivation remained constant.

**Include sesame, mushrooms, tobaccos, etc.

Source: *Nongchong Such'op, 1977* (*Handbook of Agricultural Policy*), (Seoul: Ministry of Agriculture and Fisheries, 1977), p. 35-43.

Table 6
Food Consumption, Production and Import
South Korea, 1961-1976

Year	Consumption	Production	Imports	Self-Sufficiency
1961	5,463	4,993	536	91.4%
1962	5,976	5,429	482	90.8%
1963	6,138	4,637	1,223	75.5%
1964	6,408	5,996	84	93.6%
1965	7,313	6,864	634	93.9%
1966	7,089	6,715	525	94.7%
1967	8,014	6,947	1,100	86.7%
1968	7,976	6,486	1,496	81.3%
1969	8,573	6,307	2,389	73.6%
1970	8,820	7,097	2,115	80.5%
1971	9,856	6,842	2,883	69.4%
1972	9,626	6,807	3,210	70.7%
1973	9,715	6,538	3,249	67.3%
1974	9,470	6,674	2,732	70.5%
1975	9,561	7,295	3,147	76.3%
1976	10,271	7,692	2,849	74.0%

Source: *Nongchong Such'op,* 1977, pp. 66-67.

have stimulated improved production of rice and have encouraged increases in other crops. During the first three Five Year Plans production of vegetables almost tripled to 3.2 million tons, as increased consumer purchasing power pushed up demand and prices in free markets. Fruit production multiplied four times and that of potatoes almost doubled.[12] Soya bean yields still are low, averaging the equivalent of 17 bushels per acre or one-half to one-third of what a midwestern American farmer would expect, although the prices paid Korean growers are about 4 times the world market. And South Korea now imports over one-half million metric tons of soya beans annually, plus 3.5 million tons of corn and 1.7 million tons of wheat. Feedgrain imports reflect the burgeoning demand for meat and fish as South Koreans shift to eating more high protein foods. Retail costs are relatively high: chicken sells for the equivalent of US$1.22 per pound, pork and fish are 50 percent more costly, and beef is twice the price of poultry.[13] In periodic attempts to placate consumers, the government allows major imports of pork or beef from Australia, unfortunately without sufficient price monitoring to prevent wide fluctuations in market prices that can hurt Korean farmers.

Despite some grumbling by farmers and occasional rough-handed methods by the authorities, the result to date has been the creation of a prosperous and increasingly politically conservative rural constituency in South Korea. In the elections of 1963 and 1967 President Park and his party had received substantial rural support. They lost much of this support in the 1971 elections when farmers felt their living standard had fallen behind that of urbanites. It was this widespread perception of inequity that prodded the government to undertake the mass mobilization of the countryside through the *Saemaul Undong* and to improve prices farmers received for their product. By 1974 income of the average rural household already exceeded that of urban families, an advantage which has been sustained as living costs of farmers are less vulnerable to inflation than those of city dwellers. Average annual farm family income increased about 5 times from US$747 in 1970 to US$3,893 in 1978. Farm laborers—paid at least W2,500 to W3,500 plus meals, cigarettes, and *makkoli*, the milk-colored rural alcoholic drink—today earn the equivalent of US$11 to US$13 per day. A skilled carpenter in the countryside can earn US$16 to US$20 daily.

As happened earlier in Japan and Taiwan, in South Korea a rural society composed mainly of owner-operating farmers who are prospering has become a primary foundation for stability in the country. It required far more than land reform to achieve this productivity and better living in the villages, including response by officials to what farmers saw as their vital interests. It did make a "new Confucianism," or *yushin*, which President Park defined as the national purpose, real and meaningful in the countryside. Whether this will be a sufficient "national center of gravity" since his death in October 1979 and the resulting political confusion has become a critical question for the future, not only for South Koreans but also for the international community. Two decades ago the level of expectations in South Korea was low and most families were grateful just to survive. Since then the per capita GNP has increased about 18 times to more than US$1,550 annually in 1980. At the same time South Korea has become economically hostage to what happens in world markets, from the prices paid for imported oil to sales abroad of exports such as textiles, shoes, and, increasingly, heavy equipment and other higher technology manufactures. Unemployed young people in the cities can often go home to their transformed rural villages. Yet the inputs required by a sophisticated agriculture, particularly fertilizer and fuel, also are affected by the world prices. Thus the new prosperity on the farm is more fragile than the old, predominantly subsistence pattern and far greater skill in national decision making will be required to keep it productively responsive to future opportunities.

(June 1980)

NOTES

1. *Economic History of Chosen* (Seoul: Bank of Chosen, 1921).

2. Sidney Klein, *The Pattern of Land Tenure Reform in East Asia After World War II* (New York: Bookman Associates, 1958).

3. Ibid.

4. Cornelius Osgood, *The Koreans and Their Culture* (New York: Ronald Press Company, 1951).

5. After a 92-day battle in the mountains west of the Salween River gorge near the Burma-China border in 1944, this writer interviewed the 14 survivors of an original 24-woman contingent sent to "service" a Japanese Army mixed brigade of about 3,600 men. They also were required to wash the officers' laundry. These were peasant girls recruited with offers of jobs in Japanese factories and instead placed aboard a ship that took them to Rangoon from where they were sent up the Burma Road.

6. Bruce Cumings, "Political Participation in Liberated Korea: Mobilization and Revolt in the Kyongsang Provinces, 1945-50," Vol. 1 (Seattle, Wash.: University of Washington Press, 1979).

7. The actual measure is the *chongbo*, equal to 2.45 acres. Most land measurements in Korea are expressed in *p'yong*, equal to the Japanese *tsubo* of 35.55 square feet, which is the surface of two *tatami* mats on which husband and wife sleep.

8. Klein, op. cit.

9. From 1954 through 1975 the United States devoted roughly 40 percent of its economic aid to South Korea to agriculture, for a total of US$1,799 million plus the local currency equivalent of US$239 million generated by sales of commodities. Among the largest items of foreign assistance expenditure was US$513.9 million for importation of chemical fertilizers before 1967, when construction of the Gyeongju chemical fertilizer complex started South Korea on the path toward self-sufficiency in fertilizer manufacturing capacity, although raw material must still be imported.

10. The North Koreans and Chinese Communists together may have taken casualties of over two million men killed in action and wounded plus civilians in the north inevitably caught in the fighting. "Human wave" assaults on the Red Army were particularly costly where the United Nations Forces could concentrate their immense artillery fire. Amid the turmoil of the first year of war some 600,000 additional refugees escaped from North Korea to the south. South Korean government figures indicate some 165,000 of their civilians were killed during the invasion south of the 38th parallel and another 123,000 were arrested and taken north. They are part of an estimated more than one million civilians killed, wounded, or missing in the Republic of Korea, which also suffered some 300,000 military casualties.

11. Young Whan Kihl, "Politics and Agrarian Change in South Korea: Rural Modernization by 'Induced' Mobilization," in Raymond F. Hopkins, et al., *Food, Politics, and Agricultural Development: Case Studies in the Public Policy of Rural Modernization* (Boulder, Colo.: Westview Press, 1979).

12. The rate of increase in production of foods other than rice has been somewhat lower in the 1970s than in the 1960s. Current increases in food production are nonetheless impressive and continue to outstrip population increase by a healthy margin (see Table 3).

13. A recently available survey by a Korean government bank indicated that in 1965 the poorest among the population spent 72 percent of their income on food whereas by 1977 this declined to 55 percent, while the comparable figures for the wealthiest class of Koreans was 56 percent in 1965 and 40 percent in 1977. In 1978 the average South Korean ate 124 kilograms of milled rice, 30 kilograms of wheat, and 17 kilograms of barley, while he consumed only slightly over 6 kilograms of sweet or white potatoes. The revolution under way in consumption of protein is dramatic: the average South Korean ate 114 eggs in 1978 compared to 31 in 1961. Average consumption of milk products is still only 8.6 kilograms, but it compares with .05 kilograms per year in 1961. In the same period poultry consumption has more than tripled to a current 2.4 kilograms annually, pork eaten has more than doubled to 5.1 kilograms, and beef consumption has multiplied more than 5 times from a mere one-half kilogram per capita annually in 1961 to 2.7 kilograms. Although similarly precise figures are not available for seafoods, vegetables, and fruits, it is apparent that demand has increased rapidly as rising incomes are invested primarily in better eating.

Indian woman

Andean woman

Ecuadorian Agrarian Reform: The Politics of Limited Change

Howard Handelman

On March 15, 1980, Ecuadorian President Jaime Roldós and his Vice-President Osvaldo Hurtado announced the general outlines of their new administration's five-year economic plan. The ambitious $5.4 billion proposal—calling for major reinvestment of the nation's petroleum wealth—included $250 million for rural development and the distribution of expropriated private and public land to some 77,000 needy peasant families. If such a program is actually implemented—no certainty in a country where impressive legislation is often stillborn—it would significantly expand the government's halting efforts at agrarian reform begun on July 11, 1964, when Ecuador's ruling military junta passed the first of several agrarian reform decrees and amendments. Rather than effecting radical change, however, much of this legislation reinforced socioeconomic trends which had already been modifying rural land tenancy and landlord-peasant relationships.

Ecuador has traditionally had one of the most rural populations in South America and one of the highest proportions of its work force engaged in agriculture. In 1970, the agricultural sector accounted for 56 percent of the economically active population.[1] Prior to the 1973 petroleum boom, cocoa, bananas, sugar, and coffee accounted for over 90 percent of Ecuador's export revenues, although agricultural output generally stagnated in the 1970s. That sector's share of the GDP declined from 39 percent in 1954 to 27 percent in 1970 and 21 percent in 1978. More important, per capita food production declined.[2] Despite its relatively small population, Ecuador has a very low level of cultivated land per capita.[3] Moreover, as in much of Latin America, land tenancy has traditionally been highly concentrated. Prior to the first agrarian reform legislation, some 1,400 *latifundia* (large estates) of 500 hectares (1,250 acres) or more—representing but 0.5 percent of the nation's agricultural units—controlled 45 percent of the agro-livestock land.[4]

While agrarian reform has been motivated by a number of political and economic factors, its proponents clearly have hoped that it would also contribute to greater agricultural productivity. Thus the poor production record over the past decade or more and the need for growing food imports have subjected reform proponents to severe criticism from various points on the political spectrum. Many social scientists, as well as spokesmen for peasant groups and their allies within the Catholic Church and the political party system, argue that efforts to redistribute land and to aid the small farmer have been half-hearted and inadequate in scale. Indeed, in comparison with some other Andean nations—Peru and Bolivia—land redistribution and agrarian change in Ecuador do appear quite modest.

Ecuador's larger landowners (*hacendados*) take a different view. For them, agrarian reform is the bête noire: it has proliferated inefficient smallholdings (*minifundia*) they insist, and has transferred land from the hands of allegedly efficient estate (hacienda) owners to peasants who lack the technical skills needed to farm it productively. Even enlightened landowners such as former President Galo Plaza Lasso (1948-1952), one of Ecuador's more progressive chief executives and a spokesman for the highlands' most modern hacendados, expressed similar sharp criticism to this author. "If a latifundia is being inefficiently farmed," he argued, "it should be subject to redistribution. All too often, however, the most productive estates are turned over to peasants who don't know how to farm them. As a result, output falls precipitously."

Plaza Lasso's negative evaluation is widely shared. In a series of interviews I conducted during the course of Ecuador's 1978-79 national elections, spokesmen for most of the major parties—from the Conservatives and Social Christians on the right to the Democratic Left and Christian Democrats on the left (excluding the small, Marxist splinter parties)—maintained that raising food production must be the new administration's first rural priority and that such a goal precluded significant land redistribution.[5] Indeed, the inclusion of a major agrarian reform plank in the five-year economic plan (written under the direction of Vice-President Hurtado) is intriguing, since Hurtado and a top administration economic adviser told me both before and after the 1979 election that the Roldós government was unlikely to initiate an extensive land redistribution program.[6]

This article will discuss the political and socioeconomic context of Ecuador's past agrarian reform efforts, analyze the degree of change that has resulted, evaluate the reform's successes and failures, and speculate about future developments.

The Structure of Ecuadorian Agriculture

Ecuadorian agricultural production is concentrated in two distinct geographical zones: the warm weather region stretching from the arid Pacific coast to the tropical area on the western slope of the Andes, and the colder, high altitude zone of the inter-Andean plateau (sierra). The coastal region has been the source of rice, most meat and fruit, and export crops—bananas (also a domestic staple), sugar, coffee, and cocoa. Highland agriculture, on the other hand, is dominated by ranching (dairy cattle, sheep) and the cultivation of potatoes, corn, barley, legumes, and (at lower altitudes) wheat. Sierra production is overwhelmingly for the domestic market.

Since the early 1970s some migration has taken place from the highlands and coast into the sparsely inhabited Amazonian jungle to the east of the Andes. Newly colonized lands have been used for raising cattle, corn, and fruit, and there are plans for initiating extensive palm oil cultivation. Though there is much talk among politicians and in the mass media of the *oriente*'s (eastern jungle region) potential for colonization and the resettlement of land-hungry peasants, ecological limitations would (or should) limit that option. The oriente does not play a significant role in national food production at present.

Land tenancy in both the inter-Andean and coastal regions of Ecuador has traditionally been characterized by a pattern common to most of Latin America: the concentration of a large portion of the agro-livestock land in a small number of vast estates and the proliferation of smallholdings, most of them too small to provide subsistence. Tables 1 and 2 indicate that both on the coast and in the sierra several hundred large estates representing less than one percent of the agricultural units in each area controlled over 40 percent of the agro-livestock land in the 1950s.

Important differences distinguish the two regions, however. During the 1950s the role of the medium-sized farm (10-50 hectares) was already far more important on the coast than in the highlands. The construction of new roads into the Santo Domingo-Quevado tropical region on the western slope of the Andes during the 1950s opened that area to considerable colonization during the next two decades and greatly expanded the role of medium- and large-sized units. These new holdings were tremendously important in the 1950 banana boom which brought about a 1,000 percent increase in the land devoted to that crop.[7]

The large estates both on the coast and in the highlands have featured a wide variety of landlord-peasant tenancy relationships: share cropping, cash renters, neofeudal labor payments (as a form of rent), and paid agricultural workers. In the sierra some 30 percent of the peasantry in the early 1960s rented land, and some independent smallholders (66% of the peasantry) supplemented their income with occasional hacienda employment. While perhaps 15 percent of the sierra peasantry secured their plots (or even the use of pasture and infrastructure) through virtually unpaid labor on the estate owner's land (these peasants were known as *huasipungueros, yanaperas*, etc.), such neofeudal arrangements were not common on the coast. Conversely, though the Ecuadorian coast has featured far fewer salaried plantation workers than neighboring Peru, the coastal region still traditionally has more hired agricultural workers than the highlands. Similarly, tenants on the Ecuadorian coast were far more likely than their Andean counterparts to pay cash rents for their plots. Capitalist labor and tenant relations were thus entrenched earlier and more widely on the coast than in the sierra. In the southern and

Table 1: Land Tenancy in the Ecuadorian Highlands (mid-1950s)

Size (hectares)		Units	Area (thousands of hectares)
Minifundia	0-9.9	234,596 (90.4%)	496.4 (16.4%)
Medium	10-49.9	18,292 (7.0%)	362.0 (12.0%)
Large	50-499.9	5,962 (2.3%)	689.8 (22.8%)
Latifundia	500 or more	719 (0.3%)	1,472.2 (48.8%)

Table 2: Land Tenancy on the Ecuadorian Coast (mid-1950s)

Size (hectares)		Units	Area (thousands of hectares)
Minifundia	0-0.9	53,340 (63.0%)	207.3 (7.0%)
Medium	10-49.9	27,256 (32.2%)	852.3 (28.6%)
Large	50-499.9	3,419 (4.0%)	685.0 (23.0%)
Latifundia	500 or more	650 (0.8%)	1,234.5 (41.4%)

Source: CIDA, *Ecuador: Tenencia de la Tierra y Desarrollo Socio-Económico del Sector Agrícola* (Washington, 1965), p. 522.

south central provinces of Loja, Azuay, Cañar, and Chimborazo—where the Quechua (Indian) peasantry is more isolated from the modern socioeconomic system—neofeudal relationships endured until quite recently.

The continual opening of new agricultural lands on the western slope of the Andes added to the coastal peasant and agricultural worker's greater geographical mobility. Because of lower land pressure and greater opportunities for peasants to move into newly colonized lands, coastal landlords were generally unable to exert as much control over their tenants and hired labor as in the highlands. Boom and bust cycles in coastal agriculture also led to greater fluidity in landlord-peasant relationships.

Rural Change and Limited Peasant Mobilization

Any comprehensive understanding of the various agrarian reform decrees issued during the 1960s and 1970s requires a brief description of critical changes in production, and accompanying alternations in landlord-peasant relationships, which preceded those reforms. Considered in this context, agrarian reform did not represent a sharp break with the past but was instead an extension and acceleration of existing trends.

Coastal Agriculture. Whereas the highland hacienda system can be traced back to colonial times, the concentration of coastal agriculture into large estates dates only to the cocoa boom during the last third of the nineteenth century. Patron-client relationships were never firmly entrenched in the coastal regions of Guayas, Los Rios, and El Oro. During the last years of the nineteenth century and the first decades of the twentieth, Ecuador became the world's largest cocoa exporter (contributing 35% or more of total commercial production in certain years).[8] Cocoa growers established close linkages with the Guayaquil banking system to form an ascendant plantation-banking elite. It was this group which provided the economic backing for the Liberal Party's rise to power in the early years of this century.

In 1922, however, a devastating cocoa blight hit the Ecuadorian coast and was followed by a decreasing demand in the world market for that crop. The coastal cocoa economy collapsed, and Ecuadorian exports declined sharply for the next two decades. Between 1920 and 1940 total exports dropped at a rate of 4.8 percent annually (in constant dollars) until Ecuador reached one of the lowest levels of exports per capita of any Latin American nation.[9]

As a result of these market changes, coastal agriculture was transformed. Lands previously devoted to cocoa were converted to rice, bananas, sugar, and subsistence farming. Plantation agriculture predominated only in sugar cultivation, with control lodged with the powerful San Carlos group of Guayaquil-based bankers, exporters, and growers. The same conglomerate later became the leading exporter of bananas as well.

For the most part, however, the post-1922 period featured a significant reduction in landlord control over the coastal peasantry. Many of the established estate owners sold their land to new absentee owners who lacked legitimacy vis-à-vis tenants. In many cases former tenants purchased their plots or moved into newly cleared lands to claim small land parcels. Even where the land remained in the hands of larger landlords—the more common situation—tenants who previously were forced to sharecrop or to sell their crop (particularly cocoa) to the landlords at reduced prices now began to rent their land on a fixed cash basis and to sell rice or bananas directly to the domestic market.

The transformation of many former sharecroppers into cash renters or even smallholders was not without tension. In the rich rice-growing Guayas basin of Guayas and Los Rios provinces (characterized as "perhaps the most fertile rainy tropical lowlands in the Western hemisphere"), many landlords tried to maintain traditional sharecropping arrangements, extracting 10 percent of the tenants' rice crop as rent and forcing them to sell the rest to the owners' mills at prices well below market value.[10] As the landlords expanded their holdings into previously uncultivated areas, they would try to move tenants off old lands and use them to clear the new holdings. Not surprisingly, such practices were resisted by many peasants who wished to buy their plots (so as to raise subsistence crops and market their surplus directly) or at least to rent land for cash. Even when owners agreed to rent or sell the plots to their tenants, conflicts arose over the terms.[11]

During the late 1940s and early 1950s, new developments in the world trade cycle once again produced major changes on the Ecuadorian coast. As cocoa and coffee prices rose sharply, many absentee owners tried to reassert control over their land. These moves were resented and often resisted by tenants now accustomed to paying fixed rents (independent of the volume or value of their output) and selling their crops (rice, cocoa, bananas) independently. More important, a tremendous boom in banana exports (resulting from the destruction of Central American production through hurricanes and disease) brought further efforts to reinstate the plantation system. United Fruit Company, a large Chilean firm, and Ecuadorian owners all sought to expand their holdings in the booming banana market.

Under these pressures, tenants in the Guayas valley and other coastal regions began to organize local unions. Seeking outside support, some peasant groups (particularly in the Milagro regions of Guayas) turned to the Ecuadorian Labor Confederation (CTE), an affiliate of the Ecuadorian Communist Party. During the late 1950s, as banana exports receded somewhat, the CTE also began to organize salaried plantation workers who were being laid off by United Fruit and other growers.

In 1955, the CTE led a major strike in the northern coastal province of Esmeraldas against the Swedish-owned ASTRAL banana plantations. Four years later, a violent eight-month strike was carried out against Ecuadorian- and U.S.-owned banana plantations in the Quevado region. Serious labor unrest on the United Fruit banana plantations in 1962 led that firm to sell most of its land to the Ecuadorian government. Concurrent with these organized plantation strikes, many tenants in the provinces of El Oro and Guayas were invading hacienda lands or simply refusing to pay rent on their plots.

During the 1959 presidential election rural conflict in the coastal provinces became a campaign issue as winning candidate José María Velasco Ibarra promised to enact agrarian reform legislation and to recognize the coastal plantation unions. Velasco, a demagogic populist, was elected to the presidency for the fourth time (of five) and promptly failed to fulfill his promise of rural reform. Feeling betrayed, tenants in Guayas and Los Rios carried out a series of land invasions and strikes from 1960 into 1961. In November 1961, the military ousted President Velasco for the third time and replaced him with Vice-President Carlos Arosemena. Despite Arosemena's left-of-center reputation and his professed support for agrarian reform, his administration repressed further strikes by tenants in the Guayas region. By 1963, however, internal and foreign pressures for some form of agrarian legislation could scarcely be resisted any longer.

The Sierra. In the comparatively traditional regions of the inter-Andean plateau, rural change was more gradual with less peasant mobilization. Just as fluctuations in international market conditions altered coastal tenancy and labor relations, however, changes in the domestic market had similar effects in the highlands.

While variations in agricultural and labor patterns in different regions of the sierra make it difficult, it is nevertheless possible to observe some common characteristics of highland agriculture. In every province of the sierra, for example, land tenancy prior to the 1960s was dominated by estates of 100 hectares or more. For the Andes as a whole, 64.3 percent of all agro-livestock land was concentrated in these large units. More strikingly, the 389 largest latifundia (a mere 0.2% of all highland units) controlled 41 percent of the agro-livestock land. At the other end of the spectrum, 81.7 percent of the sierra's agricultural units were under 5 hectares. Together, these 212,000 minifundia controlled only 11.4 percent of the land. Indeed, 32 percent of all sierra farms were under one hectare and accounted for only 1.4 percent of the agrarian land.[12]

The dominant latifundia featured a mixture of livestock ranching (sheep and cattle) and the cultivation of grain (corn, wheat, barley), tubers (especially potatoes), and legumes. In all, some 66 percent of the sierra peasantry (ca. 1960) owned their own small plots, but some 30-40 percent of the peasant work force (including some property owners who could not support their families on their plots) were dependent on the hacienda system. They might offer up to four or five days' labor per week on the hacendados' land for a nominal wage (or even no wage at all) in return for usufruct of a small plot of land and pasture grazing rights (*huasipungueros*). Some peasants (*yanaperos*) offered their labor merely in return for pasture rights and access to water and roads, while still others (*arrendatarios* and *partidarios*) were granted plots of some type of sharecropping arrangement. Finally, some landless peons worked for a wage only.[13]

During the 1950s and early 1960s it became increasingly obvious to the more entrepreneurial estate owners in the central and northern sierra that a growing urban middle class and skilled working class were creating a demand for certain foods—particularly meat and dairy products—that had previously been limited to a small elite. By the late 1950s over 20 percent of the population (with nearly one-third of national income) could be categorized as middle class. From 1961-1965, as population grew at 3.4 percent annually, private consumption increased annually by 5.3 percent (one of Latin America's highest rates of growth in per capita consumption for that period).[14] Traditional Andean hacienda agriculture—based on extensive land use, high input of poorly paid labor, and little capital investment—was incapable of responding to this growing market. Innovative hacendados in the provinces of Pichincha, Cotopaxi, and Imbabura (particularly the regions of Cayambe and Machachi), therefore began to modernize production methods and convert crop lands into more lucrative dairy production. Dairy farming—the most profitable form of highland agriculture for most of the past 25 years—became the leading sector in agricultural modernization.[15] In 1961 Ecuador's first pasteurization plant was constructed in Quito: 16 years later there were 22 plants, 19 of them in the sierra.

This modernization of dairy haciendas involved the improvement of breeds, increased irrigation and application of fertilizer to pastures, and, more recently, artificial insemination and mechanized milking (though the last is still rare). In the central Andean province of Cotopaxi, "artificial" (irrigated) pasture area increased by 350 percent between 1954 and 1968 and by another 67 percent in the succeeding 6 years. Daily yield per milk cow was increased on the major haciendas from under three liters to ten liters.

To the modernizing landlords, the old semifeudal labor relations were increasingly unattractive. For one thing, the traditional order obligated a hacendado to provide employment and certain services to his peasants and many of his relatives. Even though wage payment was minimal, many landlords preferred a smaller, higher-paid, full-time work force. In addition, traditional pasture rights were restricting the landlords' ability to improve cattle stock, while the usufruct of plots by sharecroppers might impinge on the consolidation of artificial pasture.

Many landlords realized that their most profitable production strategy would be to consolidate their investment (irrigation, fertilization, improved milk sheds, etc.) on a reduced quantity of higher quality land. Tenants were offered ownership of small plots on less desirable, out-of-the-way terrain. A few of the most progressive hacendados—notably former President Galo Plaza, perhaps the sierra's most modern dairy farmer—aided former tenants with artisan co-ops, loans, and technical assistance. More commonly, the new "property owners"—often relocated on inferior plots—were left to fend for themselves. In some instances tenants were simply evicted.

Between 1959 and 1964, when the first agrarian reform law was introduced, one-sixth of the sierra's peasants who claimed traditional grazing rights on estate land were transformed by their landlords into smallholders. In the more capitalistic agrarian provinces of Imbabura, Pichincha, and Carchi (in the northern Andes), the figure reached 25-40 percent.[16] In changing their status from tenants to smallholders, the peasants usually lost their traditional rights to pasture usage, water, firewood, etc. In other parts of the sierra, particularly in the south, some of the more inefficient, neofeudal hacendados sold plots to their former tenants and sharecroppers. Since those peasants generally could not earn enough on their small plots to pay off the cost of their new land, they were required to do so by rendering free labor (whereas previously they received a nominal wage and the traditional rights just mentioned). Most former hacienda tenants, especially in the south, found themselves worse off as the capitalist pattern spread.

Even under the most favorable of circumstances, the erosion of traditional peasant-landlord labor and tenancy relations tended to weaken the long-standing legitimacy of hacendado authority. In areas such as Chimborazo (where some of Ecuador's poorest, most feudal haciendas are found) open tensions developed on a number of latifundia where hacendados refused to pay even nominal wages and evicted tenants from their plots. Aided by sympathetic Catholic clergy, peasants on several estates formed unions demanding that their landlords pay them the wages required by law. In Cayambe—a modern dairy-producing region—a number of hacienda unions were formed as early as the 1940s with the support of the Federation of Ecuadorian Indians (FIE), a wing of the CTE. During the 1950s and early 1960s workers on several Cayambe haciendas successfully struck for the legally guaranteed minimum wage and paid holidays.[17]

These examples of organized peasant unionization and the more extensive rural mobilization on the coast suggest that Ecuador's peasants have not been as passive as they are frequently depicted by academic observers.[18] Yet Ecuadorian peasant mobilization—particularly in the sierra—has not approached the level of that in Peru, Bolivia, or Chile. Unionization efforts have been fairly localized and peasant participation in the political process has probably been lower than in any other Andean nation, including Colombia and Venezuela.

1962-1964: The Debate Over Agrarian Reform

With the victory of Cuba's rural-based revolution in 1959, agrarian reform became a political agenda item in most nations of Latin America. In none of these nations did the state significantly restructure land ownership, although international pressures and the fear of rural unrest induced the political elite at least to pay lip service to the issue.[19] Ecuador was no exception. The formal commitment of the United States (through the Alliance for Progress) and of the Organization of American States (through the Punta del Este Conference) suggested that continued foreign aid might require the passage of some kind of agrarian reform.

On January 31, 1960, President Velasco Ibarra appointed a National Agrarian Reform Commission to draw up potential legislation. From that point through mid-1963, under Velasco and his successor Carlos Arosemena, a variety of presidential and congressional commissions and independent spokesmen representing political parties and interest groups issued various legislative proposals.[20] As voices such as the FAO and the Ecuadorian commission to the Alliance for Progress added their support for rural change, it became clear that some kind of agrarian law would be passed.

Three questions formed the core of the political debate:

1. How much redistribution (if any) of privately owned latifundia would take place (as opposed to colonization of new areas and the distribution of state property)?

2. If limits were to be placed on the size of private estates, how large should they be?

3. What types of traditional labor relations in the countryside should be restricted or abolished?

From the outset the large landowners—organized regionally into a series of Chambers of Agriculture—influenced the decision-making process in a number of ways: through the direct role of hacendado spokesmen serving in the national Congress (such as Senator Marco Tulio González, president of the First Zone [north-central sierra] Chamber of Agriculture); direct representation on several of the presidential commissions; mass media campaigns, particularly in influential Guayaquil and Quito newspapers like *El Telegrafo* and *El Comercio.*

Ecuador's peasantry and agricultural workers had far more limited opportunities to influence decision making. Strikes by plantation workers and tenants on the coast put some indirect pressure on the government. In 1962, for example, an FTE strike at United Fruit's Tenguel plantation induced that company to sell its property to the government's National Institute of Colonization. In the Quevado region of Los Rios province and in Guayas province tenants invaded several banana and rice plantations. A large peasant march on Quito sponsored by the CTE and FIE drew serious attention from *El Comercio* which spoke of "a peaceful invasion of the city...." Shortly thereafter, President Arosemena (who had just assumed office) addressed the third national congress of the Federation of Ecuadorian Indians (FEI) in Quito.[21] Nevertheless, the fear of peasant unrest that did exist derived more from events elsewhere—the Cuban Revolution and massive land invasions in Peru—than within Ecuador.

From the start, the various sectors of the landowning elite were agreed on two issues: first, there would have to be some kind of agrarian reform legislation if for no other reason than to satisfy the international community; second, no serious redistribution of private latifundia land should take place. Beyond this, there were attitudinal differences both between the coastal and the sierra rural aristocracy and within each region.

The powerful Guayaquil Chamber of Agriculture—dominated by the San Carlos group and other large sugar, rice, and banana growers—were concerned about land ceilings since they owned vast tracts, often for future development, in the fertile region west of the Andes. Because land pressure was greater in the sierra, they feared expropriation. "The technification of agricultural production rather than political mobilization of the peasantry," they argued at a June 1963 congress, "should be the primary objective of the agrarian reform."[22] They called for colonization of the oriente and the distribution of state lands rather than expropriation of private property. Reflecting the views of the Guayaquil elite, *El Telegrafo* charged that the National Economic Council, which was drawing up the legislation, was dominated by Andean economic interests who were trying to make coastal landowners pay for the inefficiency of highland agriculture.[23] Their anxieties were increased when a July 1963 coup ousted President Arosemena and installed military officers with Andean political ties.

The more feudalistic landowners of the southern sierra also feared serious restriction on latifundia size since their inefficient, labor-intensive estates were dependent on the use of large areas and little capital investment. Indeed, they were the most reactionary sector of the landowning class since they also opposed the prohibition of certain traditional labor relations.

A more modernizing hacendado class in the central and northern inter-Andean plain was willing to accept the abolition of the tenant system and other neofeudal peasant-landlord bonds. Indeed, as noted, most of them were already voluntarily converting their tenants to smallholders or wage-earning workers. Many of these landlords were still not sufficiently efficient to reduce their holdings significantly, however, and hence opposed meaningful legislative limits on maximum holdings.

Finally, the most modern sierra dairy producers (represented by Galo Plaza and the Cayambe and Machachi landlords) were amenable both to the abolition of feudalistic labor relations and to some limitation of hacienda size.

Decree 1480: The Agrarian Reform Law

On July 11, 1964, Ecuador's ruling military junta finally issued an agrarian reform decree after four years of national political debate. The legislation permitted coastal landowners to retain up to 3,500 hectares of crop, pasture, and "reserve" land and exempted holdings up to 1,800 hectares in the sierra. These maximums were sufficiently large to satisfy both the coastal elite and the modernizing Andean hacendados. Moreover, the law was replete with loopholes. Lands producing crops which were "destined for industrial transformation" were specifically excluded from agrarian reform provisions, thereby exempting the powerful coastal sugar growers. Corporate (as opposed to individually owned) land could also be exempted, another loophole widely used on the coast. The decree further excluded any latifundia land "efficiently exploited"—a term sufficiently vague so as to encompass any modern hacendado or company with political influence. Indeed, IERAC (the newly created Ecuadorian Institute of Agrarian Reform and Colonization), the organism charged with administering the reform legislation, could extend maximum allowable land ownership in any situation which it deemed proper. Finally, first priority was to be placed on the distribution of government-owned lands. In short, the limits on private land ownership were virtually meaningless, their potential effect limited to the most inefficient, traditional southern Andean estates.

The law did abolish the most feudalistic highland labor relations (*huasipungo* and *yanapa*) which involved the peasants' payment of rent (for land, pasture rights, water rights, etc.) through virtually unpaid labor. (Ecuador was one of the last nations in South America to permit that practice and it had become an embarrassment.) Former tenants were to be given title to their small plots with their payment to the landlord (if any) depending on the length of time that they had worked on the hacienda.[24] They thus forfeited their traditional rights to hacienda water, firewood, access road, and—after five years—pasture usage. The law further mandated the abolition within eight years of certain sharecropping relations still practiced on the more traditional coastal cocoa estates. However, sharecropping on rice lands (a far more important crop) was specifically permitted to continue.

The military government and IERAC stated that the intent of the agrarian reform was to encourage productivity, transfer inefficient latifundia lands directly to the peasantry, and "integrate" the smallholder into the national economy. In order to facilitate the last goal, preference in land redistribution was to be given to peasant cooperatives under the jurisdiction of IERAC.

By virtually any criterion, however, the 1964 law was weak. The more feudalistic types of landlord tenant relations it prohibited were already being abandoned in much of the inter-Andean plain and were not of consequence for most of the coast. By the start of the 1960s tenants accounted for only 7 percent or so of the highland peasantry and sharecroppers for only 3 percent of the Andean peasant population and a comparably small minority on the coast.[25]

For the most part, the basic structure of land tenancy was kept intact. Given the great political influence of the most capitalistic landowners and the weakness of peasant organizations, such an outcome was not surprising. Indeed, the agrarian reform clearly demonstrated the ability of major banana and sugar growers and of the modern Andean dairy farmers to protect their interests. Such reforms as were actually implemented sacrificed the most inefficient and politically ineffective landlords—the southern Andean hacendados and the coastal cocoa growers—to the winds of change.

Aftermath in the Andean Highlands

Between 1964 and 1971, 17,468 Andean tenant families were given title to 60,472 hectares of land (over 85% of that transfer taking place in the years 1965-1967), thereby practically eliminating tenancy relationships.[26] The reform had the short-term effect of increasing landlord-peasant conflict in some regions and of inspiring further peasant mobilization. In many parts of the southern inter-Andean plateau, hacendados anxious to avoid loss of their tenants' plots began to evict the tenants or to convert them to salaried workers. Several landlords in the provinces of Chimborazo and Azuay told this author they were able to evade transfer of peasant parcels in this manner. In a few regions of the southern and central sierra, peasants organized local unions (*sindicatos*) to prevent such evictions. In other cases, tenants working under arrangements still permitted by law (*aparceros*) organized for the purpose of securing their plots. During the late 1960s and early 1970s peasant federations were formed in the provinces of Chimborazo, Tungurahua, Cañar, and Bolivar, but only in Chimborazo did peasant mobilization reach significant levels.[27] While CTE, FIE, and other leftist labor groups were of some help to the Chimborazo peasants, sympathetic clergymen, especially Archbishop Proaño (of the city of Riobamba), one of the most politically progressive and outspoken Catholic leaders in Ecuador and a major voice throughout Latin America for the theology of liberation, were more influential. A number of haciendas in Chimborazo and elsewhere in the southern highlands were expropriated by IERAC and turned into peasant cooperatives.

The most important land transfer, however, affected estates belonging to the Ecuadorian government's Asistencia Pública (Public Assistance)—the organism which funds public hospitals and other social services. During the early part of the twentieth century, the government had expropriated many church-owned haciendas and turned them over to the Asistencia Pública. As of 1960, it accounted for fully 20 percent of all sierra hacienda property in units of over 100 hectares.[28] More important, these estates frequently contained the highest quality lands in the region.

Since the early part of the century the government had leased those haciendas to large landowners. Because the hacendados did not have title to the lands and because management of the property might change periodically (as land was transferred from one major leasee to another), hacendado control over the estates' peasantry was limited. Consequently, peasant pressure was strongest here and it was these Asistencia Pública estates which provided most of the land for the cooperatives created by IERAC from the late 1960s through the mid-1970s, thereby reducing pressure for expropriation of private property.

Continued Coastal Tensions and Decree 1001 (1965-1970)

The effects of the first agrarian reform law were far more limited on the coast than in the sierra. IERAC did expropriate a number of cocoa estates belonging to politically weak absentee landlords. In addition, the Institute granted title to some squatters who had occupied property in the tropical Santo Domingo region west of the Andes. In early 1965, some land was even expropriated from the San Carlos sugar company. The military government's appointment of José Arey Marín as Agriculture Minister in June 1965, less than a year after the passage of Decree 1480, guaranteed, however, that major coastal estates would not be touched. Arey Marín was an officer of the Guayaquil Chamber of Agriculture and a brother-in-law of Carlos Luis Plaza Danín, a leading banana plantation owner and spokesmen for the coastal opposition to the agrarian reform. From 1964-1971, only 27 percent of land adjucicated to peasants through agrarian reform was on the coast.

In March 1966 the Executive Director of IERAC, Juan Casals, resigned charging "betrayal of the agrarian reform."[29] Shortly thereafter, Ecuador's military government stepped down to be replaced briefly by interim civilian President Clemente Yeravi and then by Otto Arosemena, who was elected to a two-year term (1966-1968) by a national constituent assembly. Both Yeravi and Arosemena were closely tied to the Guayaquil agriculture/export elite. In 1966 the IERAC budget was cut 40 percent and President Arosemena removed its remaining progressive officials. His conservative administration also permitted many coastal landlords to expel tenant farmers from their sugar and rice lands so as to avoid any future possibility of redistribution of those plots.

Tensions between landowners and tenants were particularly acute in the Guayas basin rice estates. Formerly a cocoa-producing region, the area had been transformed in the 1930s to Ecuador's principal source of rice—largely grown by sharecroppers.[30] During the 1950s and 1960s rice acreage in Guayas and Los Rios province increased by some 600 percent through the colonization of new lands and the transformation of mixed (cattle-rice) estates to pure rice cultivation. While total production increased gradually, however, yield per acre was dropping. Government technical advisers as well as United States AID representatives concluded that landlords were not running the estates efficiently.

In 1967-68, a severe drought in the Guayas basin intensified tenant-landlord conflicts and induced some sharecroppers to seize their plots. More important from the national government's perspective, the drought reduced rice production in Guayas and Los Rios (1966-1968) from 189,170 to 82,601 metric tons and necessitated major importation. The crisis added strength to the government's contention that existing tenancy and production relationships were inadequate.

In December 1970, President Velasco (returned to office for his fifth term) issued Decree 1001 banning sharecropping in coastal rice areas. These estates were declared "of public utility" and subject to expropriation.[31] Subsequently, leasing of agricultural land of any kind was declared illegal. Soon after the passage of the 1970 Decree, rice-growing sharecroppers began to terminate their rental payments even on unexpropriated lands. By 1976, most coastal rice lands had been expropriated by IERAC, but the agency had been rather slow in delivering title to the peasants. With the assistance of AID, the government helped create nearly 100 producer cooperatives of former rice sharecroppers.

Following the 1973 petroleum boom, Ecuador's new military regime increased the rate of rice land expropriation and adjudication and increased government loans to that sector. Credit was generally predicated on the formation of government-directed producer cooperatives (*empresas agrícolas*). Most peasants not belonging to rice-producer cooperatives nevertheless organized marketing co-ops. Government influence dominated both types of cooperative through control over rice prices (purchasing over 70% of the rice crop) and thus the state replaced the hacendado as decision-maker in rice cultivation and milling. The former sharecroppers, however, seem to have benefited economically from the transformation to cooperatives and the state has thereby effectively reduced peasant discontent and independent mobilization.

1972-1979: Agrarian Reform under Renewed Military Rule

In February 1972, the Ecuadorian armed forces once again ousted President Velasco Ibarra and returned the nation to military rule after six years of provisional and elected civilian government.[32] The new junta, headed by General Guillermo Rodríguez Lara, projected a left-of-center image somewhat modeled after the military regime in neighboring Peru.[33] Soon after taking power the government announced plans for new agrarian reform legislation which would integrate land redistribution into a broader national economic development program. Agrarian reform, it declared, should be part of "an orderly change in the economic, political, and cultural aspects of the rural structure."

The new Agrarian Reform Law issued in October 1973, superseded all previous legislation, although it offered no major innovations. Through intense lobbying, the Pichincha (Quito) and Guayas (Guayaquil) Chambers of Agriculture had watered down the proposed legislation, just as they had done in the early 1960s.[34] Internal divisions within the military, pitting more conservative officers against the Rodríguez Lara faction (and reflected in an unsuccessful September 1975 coup attempt and Rodríguez's forced resignation in January 1976), undercut the government's ability to resist these pressures.

Under the new law, the Institute of Agrarian Reform (IERAC) was changed from an autonomous agency to a division of the Ministry of Agriculture. Calling for more "preparation" of newly expropriated or colonized lands (i.e., creation of infrastructure, technical aid, etc.) before plots were turned over to peasant owners, the decree also promised additional credits for beneficiaries. Finally, the legislation streamlined provisions for adjudication of expropriated property.

To be sure, Ecuador's post-1973 petroleum boom did bring about a sharp increase in government agricultural expenditures and credits (as it did in all sectors of the budget). Between 1973 and 1974, credits extended by the National Development Bank (the major source of government agricultural loans) increased from $53 million to $133. Similarly, in 1975 and 1976, IERAC's budget was nearly tripled (in noninflated currency) from 1973-74 levels. Agricultural credits since 1975 have nevertheless failed to keep up with inflation and in 1977 IERAC's budget was slashed by over 55 percent, reducing it to near 1973-74 levels.[35]

Beginning in 1975, the military government did begin to accelerate the adjudication of previously expropriated property so that titles could be transferred from IERAC to the peasantry (Table 3). The amounts of land adjudicated in 1976 (60,436 hectares) and 1977 (75,210 hectares) were the highest annual totals since the passage of the first agrarian reform legislation in 1964. This reflected the title transfer of property previously acquired by IERAC, however, not an expansion of land redistribution. Indeed, the Ministry of Agriculture under both the Rodríguez Lara government (1972-1975) and the succeeding junta (1976-1979) emphasized colonization of new lands, particularly in the *oriente.* As Table 3 indicates, during the first 6 years of Ecuador's most recent military rule, 713,115 hectares of land were colonized (416,732 of it in the oriente), a 45 percent increase beyond the 1964-1971 period.

In December 1977, the centrist junta headed by Admiral Alfredo Povedo issued Agrarian Reform and Colonization Decree 2092. The law reflected the regime's preference for colonization over redistribution and in January 1978, the government created a new agency, INCRAE, to administer such colonization, thus depriving IERAC of a major function and undermining its authority. Peasant organizations, aware that IERAC bureaucrats, whatever their faults— were relatively more sympathetic to land redistribution, rallied to the beleaguered institute's defense. The peasants had little political influence, however, and in March 1978, IERAC director Manuel Antonio Franco resigned, citing the lack of government support for land reform. One year later, the junta issued an "Agricultural Promotion Law" which seeks to stimulate production on existing farm units (with little concern for distribution issues). Hailed by the Quito Chamber of Agriculture, the law was denounced by most peasant organizations as "an instrument at the service of large landowners."[36]

The Agrarian Reform Evaluated: Changes in Land Tenure

In 1978, the Ministry of Agriculture and the National Planning Agency (JUNAPLA) issued an extensive evaluation of agrarian reform from 1964-1976. Together with additional data published subsequently by IERAC and other government agencies, the study shows the extent and limits of changes in rural land tenancy.[37]

Table 3: Area of Land Adjudicated through Agrarian Reform and Colonization: 1964-1977

Period	Agrarian Reform			Colonization			
	Sierra	Coast	Total	Sierra	Coast	Oriente	Total
1964-1966 (Military government)	68,447	17,154	85,601	116,500	43,444	47,273	207,266
1967-1971 (Civilian government)	72,129	21,102	93,291	118,901	104,165	63,881	286,947
1972-1977 (Military government)	171,018	80,650	251,668	132,935	163,448	416,732	713,115
Total	311,594	118,906	430,500	368,386	311,057	527,886	1,207,329

Source: IERAC data in Gustavo Cosse, "Reflexiones Acerca del Estado: El Proceso Político y la Política Agraria en el Caso Ecuatoriano, 1964-1977," mimeograph (Quito: FLACSO, 1978).

Table 4: Agrarian Reform: 1964-1977

		Percentage of Agro-Livestock Area
Total Agro-Livestock Area (Sierra + Coast) (1974)	6,950,700 hectares	100.0
Cases Opened by IERAC (through 1976), of which	1,113,460 hectares	16.0
Cases Resolved (through 1976), of which	850,124 hectares	12.2
Land Adjudicated (through 1977)	427,734 hectares	6.1

Source: *Resumen del Informe,* p. 53; 1977 IERAC data; and calculations by this author.

When the first agrarian reform legislation was passed in 1964, it was announced that some 1.9 million hectares would be distributed to 171,000 families by 1976 (when the reform was to have been completed). In fact, by that date only 850,124 hectares had been acquired by IERAC, with cases pending involving an additional 240,000 hectares (see Table 4). Even were all pending cases settled, only 16 percent of the agro-livestock land in the coastal and highland regions (1,113,460 of 6,950,700 hectares) will have been affected. Through 1977, titles to only 427,734 hectares (6.1% of all agricultural land) had been officially transferred to peasant families (perhaps another 70,000 hectares were adjudicated in 1978), with the remaining 50 percent of all agrarian reform land still technically owned by IERAC (though actually most of it was in the hands of peasants). Of the 171,000 families originally promised land, only 63,757 had been affected as of 1976 and only 85,435 in all will have benefited when pending cases are completed.

In sum, all agrarian reform efforts through 1979 have touched less than 20 percent of the peasant population and less than 15 percent of the agricultural land. Ecuador's agrarian elite has been further shielded by the fact that 20 percent of agrarian reform lands (170,000 hectares) had formerly been publicly owned property and another 15-20 percent involved the transfer of huasipunguero or sharecropping plots (generally lower quality lands, with the exception of coastal rice). Table 3 reveals that the coastal banana and sugar barons have been particularly successful in protecting their interests. Through 1977, only 25 percent of all adjudicated property was on the coast, though that area actually has 30 percent *more* agricultural land than the sierra. When all currently enacted agrarian reform has been completed, less than 7 percent of the 3.85 million hectares of coastal agricultural land will have been affected.

Land Tenancy Today

While the scope of land redistribution through the agrarian reform has obviously been quite modest, this does not mean that important changes have not taken place in the structure of ownership during the past two decades. Unfortunately, there is no comprehensive data on landholding immediately prior to the 1964 agrarian reform or currently. The agricultural censuses of 1954 and 1974, however, provide important insights into changes that took place in the decade preceding the first reform and in the ten years following it.

Tables 5 and 6 reveal very similar modifications of Andean and coastal land tenancy between 1954 and 1974. The most dramatic change was the diminished role of the largest latifundia. The percentage of agricultural land controlled by estates of over 2,500 hectares declined in both regions to one-half their previous levels. Yet, as of 1974, units of 100 hectares or more (representing less than 2% of the nation's farms) still accounted for over half of all agro-livestock land. Moreover, to the extent that latifundia dominance had diminished, the major beneficiaries were the medium-sized farmers with 10-100 hectares. The absolute number of minifundia (units under 10 hectares) increased marginally in the sierra but more than doubled on the coast, where some previously landless peasants had received plots either through agrarian reform or, more commonly, through colonization of new areas (Table 6 shows a more than 25% increase [800,000 hectares] in total coastal agricultural land). The *percentage* of property owned by these smallholders only increased very modestly, however, and the average size of their minifundia actually declined from 3.9 hectares to 2.9 on the coast and from 2.1 hectares to 2.0 in the sierra. Indeed, as of 1974, 35 percent of *all* Andean farms and 18 percent of all coastal units were one hectare or less. Units that small are clearly insufficient to support peasant families in Ecuador. The agrarian reform had not produced, as of 1974, a substantial class of smallholders with units large enough to be economically viable—most analysts recommend a five-hectare minimum. The limited data available for the 1974-1979 period indicate further adjudication of land to smallholders but with no appreciable change in the pattern just described.

Table 5: Land Tenure in the Sierra -- Evolution from 1954-1974

	1954				1974			
Size of Unit	No. of Units		Area in hectares		No. of Units		Area in hectares	
0-9.9 hectares	234,596	(90.4%)	496,400	(16.5%)	280,974	(87.1%)	560,800	(18.3%)
10-19.9 hectares	10,570	(4.0%)	142,000	(4.7%)	18,266	(5.7%)	241,266	(7.8%)
20-99.9 hectares	11,316	(4.4%)	428,700	(14.5%)	19,812	(6.1%)	789,909	(25.7%)
100-499.9 hectares	2,368	(0.9%)	471,100	(15.6%)	2,935	(0.9%)	504,702	(16.4%)
500-2,500 hectares	581	(0.2%)	591,500	(19.5%)	513	(0.2%)	506,583	(16.5%)
2,500 or more hectares	138	(0.05%)	881,200	(29.2%)	86	(0.02%)	471,054	(15.3%)
Total	259,569	(100%)	3,020,400	(100%)	322,586	(100%)	3,074,274	(100%)

Source: *Resumen del Informe*; Censo Agropecuario Nacional (1954 and 1974) and calculations by CEPLAES in unpublished studies.

Table 6: Land Tenure on the Coast -- Evolution from 1954-1974

	1954		1974	
Size of Unit	No. of Units	Area in hectares	No. of Units	Area in hectares
0-0.9 hectares	53,340 (63.0%)	207,300 (7.0%)	114,928 (67.4%)	334,817 (8.9%)
10-19.9 hectares	10,830 (12.8%)	152,300 (5.1%)	20,229 (11.9%)	277,244 (7.4%)
20-99.9 hectares	16,426 (19.4%)	700,000 (23.5%)	30,111 (17.7%)	1,207,857 (32.2%)
100-499.9 hectares	3,419 (4.0%)	685,200 (23.0%)	4,990 (2.9%)	913,861 (24.3%)
500-2,500 hectares	547 (0.6%)	558,000 (18.7%)	655 (0.4%)	596,179 (15.9%)
2,500 or more hectares	103 (0.1%)	676,500 (22.7%)	79 (0.04%)	424,838 (11.3%)
Total	84,665 (100%)	2,979,300 (100%)	170,592 (100%)	3,754,796 (100%)

Source: *Resumen del Informe;* Censo Agropecuario Nacional (1954 and 1974) and calculations by CEPLAES in unpublished studies.

The breakup of many of the largest estates (over 1,000 hectares and, especially, over 2,500) into 50-500 hectare farms is attributable to several factors: the sierra trend toward more capital-intensive farming; subdivision through inheritance; division among relatives as a means of evading the agrarian reform laws; and, some transfer of sharecropper and huasipunguero plots on haciendas to their tenants.

For the roughly 20 percent of Ecuador's peasantry that had been affected by the agrarian reform through 1979, the extent of actual benefit is much debated. Many scholars argue that former hacienda peons have generally lost their neofeudal benefits, receiving title to often inferior-quality plots which in any case are too small to allow them to compete in the commercial market. The result, such critics argue, has been pauperization and marginalization forcing these peasants either to migrate to the cities or to become a rural subproletariat. [38]

My own interviews with sierra peasants suggest that the income levels of many former tenants have improved in the past five years due to factors only marginally related to the agrarian reform. Ecuador's new petroleum wealth has created a significant number of jobs—particularly in construction—on the oil pipelines and in Quito (which has experienced an oil-funded boom in middle and upper class apartment-house construction). Relieved of their former obligations to the hacendados, many ex-tenants are now free to take these jobs while their wives and children maintain their subsistence plots. On the coast, former rice sharecroppers (whose holdings average nearly ten hectares—far larger than in the sierra) have benefited more directly from the agrarian reform. In the absence of effective political pressure, however, such gains have been limited and confined to a small minority. Most Ecuadorian peasants continue to farm excessively small plots and to receive artificially low prices for their surplus crops (due to government controls designed to provide the politically more volatile cities with cheaper food).

Food Shortages and the Future of Agrarian Reform

Ultimately, the case for or against further agrarian reform must be evaluated in the context of the nation's food production and nutritional deficiencies. United Nations (FAO and ECLA) data over the past two decades reveal that Ecuador has one of the lowest average caloric and protein consumption levels of any country in Latin America, if not the entire Third World (Table 7). [39]

Undoubtedly increased purchasing power and food imports—both fueled by petroleum revenues—have improved Ecuador's *average* food consumption in the 1970s and its relative ranking within Latin America. Still, the distribution of gains has been very unequal and there is little evidence that the serious malnutrition afflicting more than 50 percent of the population has been alleviated. For urban workers, wages have barely kept up with inflation and in 1975 a typical Quito working class family spent 80 percent of its income on food! [40]

Since the early 1960s, food production has either kept just slightly ahead of population growth (1964-1970) or lagged behind (1970-1979). During the first decade of the agrarian reform programs, total food production rose 39 percent (3.6% annually), but population rose at the same rate leaving per capita consumption unchanged.[41] Since 1974, meat and dairy production have generally increased by 4-6 percent annually, but food crop cultivation has declined (especially in 1976-1979). The situation has been aggravated by serious drought in the sierra and parts of the coast from 1976-1978. During that period rainfall averaged 50-75 percent below normal in the southern and central highlands, the lowest levels in 50 years.

Production of most food crops was actually lower in 1977 than in 1970, with bananas (the major contributor to increased fruit production) and rice (a major lower class staple) among the only bright spots. Total agricultural prodution rose 2.8 percent in 1978 (double the 1977 rate but still below population growth) with gains primarily in export crops and livestock (and stagnation or decline in domestic food crops. Soft corn and potato production (both critical for the income and diet of highland peasants) dropped to one half their 1977 level and less than one-third of 1970 production! Milk output in 1978 was also some 25 percent below the previous year. [42] Faced with stagnant production (and an increased capacity to import food, using petroleum revenues), Ecuador's food imports (primarily wheat) rose from $19.6 million in 1972 to nearly $100 million in 1978.

Periodic official statements by Ecuador's Chambers of Agriculture—representing large landowners—blame food production deficiencies on the agrarian reform, poor government planning, inadequate price supports and (more recently) the drought—anything but their own production techniques.[43] In one of the more extreme attacks by the political-economic elite on agrarian reform, the editor of Ecuador's most widely circulated news magazine attributed the current "agricultural crisis" to the transfer of land to highland Indians "whose cultural level and biological condition make them totally unsuited for all but their routine, traditional [agricultural] techniques."[44] Even some relatively progressive political leaders such as Rodrigo Borja, 1978 presidential candidate of the Democratic Left, and former President Galo Plaza have insisted to this author that agrarian reform has often reduced production by transferring high quality (Asistencia Pública) lands from large landlords (who rented them) to "less efficient" peasants.

Table 7: The Americas: Average Daily Per Capita Calorie and Protein Supplies

	Calories		Protein (grams)	
Country	1961	1970	1961	1970
United States	3,120	3,270	92	97
Argentina	3,060	3,150	98	99
Paraguay	2,580	2,800	76	74
Brazil	2,430	2,600	61	64
Mexico	2,500	2,560	65	65
Venezuela	2,230	2,460	59	62
Peru	2,290	2,310	61	62
Colombia	2,180	2,250	50	51
Guatemala	1,890	2,120	53	59
ECUADOR	1,850	2,040	46	49
Haiti	1,820	1,720	40	39

Source: FAO, *The State of Food and Agriculture* (1974) cited in USDA-Economic Research Service, *Agriculture in the Americas* (Washington, 1976), Table 78.

Table 8: Food Crop Production 1970-1977

	1970	1976	1977
		(in tonnage)	
Grains and Cereals	586,419	658,055	545,050
Roots and Tubers	818,508	897,754	647,511
Vegetables (green)	160,457	175,476	136,589
Fruits	3,707,250	3,653,683	4,276,582
(Rice)	96,031	198,663	206,402
(Wheat)	81,000	65,000	39,800
(Potatoes)	541,970	499,000	417,000

Source: *El Comercio* (Quito: August 13, 1978; February 24, 1979).

Organized landowners have thus been able to defend their continued dominance of rural tenancy by arguing that land redistribution would threaten the nutritional level of the more than 50 percent of Ecuador's population not engaged in agriculture. As we have seen, they have successfully deflected government policy in recent years (e.g., the 1979 Agricultural Development Law) away from redistributive issues toward

Table 9: Size of Agricultural Units and Value of Production (1968)

Size of Unit	Total Area (hectares)	Value of Production (thousands of sucres)
5 hectares or less	706,231 (10%)	800,811 (17%)
5-50 hectares	1,888,127 (28%)	1,669,883 (36%)
over 50 hectares	3,987,775 (62%)	2,171,219 (47%)

Source: 1968 survey by the OAS and the Ecuadorian Agriculture Ministry reprinted in Fausto Jordan "La Agricultura en el Ecuador," in G. Drekonja et al., *Ecuador Hoy* (Bogotá: Siglo XXI, 1978), p. 280.

Table 10: Value of Production per Hectare by Size of Unit (1958): Coast and Sierra

(Value of Yield per Hectare in sucres)

	under 10 hectares	10-49.9 hectares	50-499.9 and over	500 hectares
Coast	2,900	2,200	1,300	900
Sierra	2,500	2,200	1,400	500

Source: CIDA: *Ecuador: Tenencia de la Tierra y Desarrollo Socio-Económico del Sector Agrícola* (Washington, 1965), p. 529.

the stimulation of production on existing large farms, with the use of colonization and an escape valve for peasant land pressures.

Given that only 16 percent of Ecuador's farmland has been affected by agrarian reform, however, it seems hard to attribute production problems to past land redistribution. Undoubtedly some IERAC-directed cooperatives on former Asistencia Pública property are now producing below their previous levels of output, but they account for only 16 percent (164,767 hectares) of the area affected by the agrarian reform and less than 2.5 percent of all agricultural land in the sierra and coast.

Indeed, if food crop production is stagnating or declining in Ecuador, it is clearly attributable to a price structure favoring the production of livestock and export crops over domestically consumed food crops and to the poor productivity record of many large landowners. In the Andean highlands, substantial amounts of land have been taken out of crop production over the past decades and converted to dairy pasture. While increased milk output has probably benefited many lower class Ecuadorians (at least in the cities), the most profitable dairy operations produce cheese which only the middle and upper classes can afford. On the coast and in the tropical regions west of the Andes (Santo Domingo), most land is devoted to meat production and to coffee, cocoa, sugar, and banana exports.

The contention that Ecuador's largest farms are more efficient than small peasant holdings fails to stand up to careful scrutiny. A 1968 OAS study indicated that productivity per hectare was *negatively* related to the size of the agricultural unit. Thus, Table 9 shows that minifundia (5 hectares or less) accounted for only 10 percent of the nation's agricultural land but produced nearly twice that proportion (17%) of its agricultural production value. Conversely, larger units (over 50 hectares) controlled 62 percent of agro-livestock land, but produced only 47 percent of the value of output. These figures are consistent with data collected in the late 1950s by the Inter-American Committee for Agricultural Development (CIDA) as well as parallel data from several other Latin American nations. As farm size increased, yield per hectare fell.[45]

The higher rate of productivity among smaller units is related to their more intensive use of the land. While farmers with units of 50 hectares or more use only 63 percent of their land for pasture or cultivation (the rest is unproductive), peasants owning units of 5 hectares or less use over 90 percent of their land.[46] In short, while there are many highly efficient dairy haciendas in the north-central sierra and similarly productive commercial plantations on the coast, most of the latifundia are either highly inefficient or fail to exploit much of their property. Indeed, when one considers the tremendous advantage large landowners have over peasant cultivators in terms of access to commercial and government loans and to state technical aid, their poor productivity record is even more striking. From 1972-1976, the Ecuadorian government loaned a total of 1.54 billion sucres ($60 million) to meat and dairy farmers. Virtually all the loans went to fewer than 400 large producers. At the same time, a 1976 survey of 50 smallholders outside Quito showed that none of them had ever received government credit or technical assistance.[47] Only recently has the Central Bank (one of the government's most progressive institutions) begun a program of low interest loans to peasants with annual incomes of under 7,500 sucres ($280).

Despite their disadvantages, many Ecuadorian peasants have shown themselves more than willing to respond to technical innovations when given the opportunity. Over 90 percent of the nation's potatoes—half of which are produced by peasants—are grown from new high-yield seeds developed by INIAP, the government's agricultural research division. Indeed, over the past five years potatoes have shown the highest increases in yield per acre of any Ecuadorian food crop. Similarly, rice—now largely a medium and smallholder crop—has had the best production record of any food staple in the 1970s.

Finally, while significant portions of land in units of 50 hectares or more are devoted to livestock (37% of the land in such units are in pasture versus 5% of the land in plots under 5 hectares), and smallholdings are far more oriented to crop cultivation (89% of minifundia land is crop cultivated versus 50% of landholdings of over 50 hectares). Though Ecuador's increased milk production has improved average urban nutritional levels, the widescale transfer of croplands into pasture has largely increased production of foods (meat, dairy products) for the middle class at the expense of lower-class staples (potatoes, corn, and cereals).[48] Moreover, through much of the inter-Andean plateau, the rich valley basins (most suited for crop cultivation) are dominated by hacienda pastures while smallholders grow food crops on mountainsides more suited for grazing.

In short, Ecuador still needs a meaningful agrarian reform. It could increase the income of many peasants currently lacking sufficient land to support their families (thereby benefiting the poorest strata of Ecuadorian society); increase the portion of rural land in production; transfer some land from pasture to the cultivation of tubers, cereals, and other food crops which are of more nutritional import to the lower classes; and increase total food production.

The mere redistribution of latifundia lands would not in itself achieve these results. Agrarian reforms of varying magnitudes in Peru, Bolivia, Venezuela, and Mexico have shown the inadequacy of land redistribution without accompanying support programs. Low-interest government (or commercial) credit and technical assistance—now unavailable to most peasants—would have to be substantially extended. Agricultural education programs would be needed to familiarize peasants with technical innovations. Government extension workers would have to overcome their present aversion to working with peasants and to dirtying their boots with soil or manure by leaving their offices for the fields. Finally, peasants would have to receive a high enough price for their crops to make their holdings economically viable and to stimulate increased production. Government price policy should also encourage production of crops most likely to reduce the caloric deficiencies of the poor.[49]

Agrarian reform will not be a panacea for rural poverty—there is not enough productive land for all peasants needing it—but it is a step in the right direction. The five-year economic plan announced by the Roldós administration promised to double the present number of agrarian reform beneficiaries, but the entire plan has been stymied in a political deadlock between the President and the Congress. Moreover, the plan leaves many important questions unanswered. How much land would be expropriated from private holdings and how much would come from the colonization of public lands, mostly in the oriente? Settlement of public land has long been politically expedient, allowing the government to avoid confrontations with powerful landed interests, but ecological restraints make wholesale colonization of the jungle an unappealing solution. One of Ecuador's most respected foreign technical advisers on tropical agriculture (with many years of experience in the oriente) suggested that even under the most optimal technical conditions (interspersing cleared pasture with nitrogen-fixing trees and legumes), the Amazonian jungle region could only resettle less than half the highland peasants who currently need land.

In the final analysis prospects for change will depend on the balance of interest group and class power within the Ecuadorian political system. Given the peasantry's lack of political influence and effective organization, there is little immediate prospect for substantial land redistribution, increased smallholder credit and technical assistance, or better crop prices. Sociologist Gustavo Cosse notes that, between 1964 and 1979, the annual rate of adjudication of agrarian reform land to peasants was greater under all three military juntas than under any of the three civilian administrations.[50] Well-organized and well-connected landowner groups like the Guayas and Pichincha Chambers of Agriculture have been able to dilute all previous attempts at reform and seem to have been most effective under civilian regimes.

In a series of interviews conducted with current leaders of the three largest congressional delegations—the CFP (a populist, urban-based party with 31 of 69 congressional seats), Democratic Left (left-of-center urban-based with 15 seats); Conservatives (rightist, landlord-dominated with 9 seats)—I found little sentiment for land redistribution or for higher crop prices. The CFP and Democratic Left (respectively centered in Guayaquil and Quito) endorse cheap food policies and show only secondary interest in peasant support. Future enfranchisement of the nation's illiterates may change this political equation somewhat, but currently the limited peasant vote is no match for either urban voter blocs or the well-organized landed elite.[51]

(March 1980)

NOTES

1. Theodore Van der Pluijm, "Evaluation of Recent Developments in Ecuador's Agriculture," Mimeograph (Rome: FAO-CIAP, 1972), p. 10. In 1974, 58 percent of Ecuador's total population and 49 percent of the economically active population was rural. See Alex Barril, "Modernización Agropecuaria y Economías Campesinas," in Barril et al., *Ecuador: Tecnología Agropecuaria y Economías Campesinas* (Quito: Brethren Unida-CEPLAES, 1978), p. 72.

2. *Latin American Economic Report* (London: October 10, 1977), p. 166; and data from Ecuador's Banco Central.

3. *Latin American Economic Report* (July 15, 1977), p. 107. There is considerable difference of opinion over the proportion of Ecuador's land that is potentially arable or cultivable. Published data on the amount of arable land per capita range from .20 to .50 hectares.

4. *Censo Agropecuario* (Quito, 1954); data compiled by CEPLAES and FLACSO of Quito.

5. National elections for the presidency and Congress were held in April 1979. In August 1979 a new civilian administration was sworn in ending over seven years of military rule.

6. Since taking office, President Roldós has come into open conflict with the Congressional majority coalition of the CFP and Conservative parties headed by CFP strongman Assad Bucaram, Roldós' former political patron and his wife's uncle. In early 1980, Roldós was expelled from the CFP, the populist party whose ticket he headed in the 1979 election. Thus, as the administration has turned more to independents and to the left-of-center DP/UDC (Vice-President Hurtado's party), it may have developed a more progressive stance on agrarian reform.

For a more extensive discussion of Ecuadorian politics, see: John D. Martz, *Ecuador: Conflicting Political Culture and the Quest for Progress* (Boston, Mass.: Allyn and Bacon, 1972); Osvaldo Hurtado, *Political Power in Ecuador* (Albuquerque, N.M.: University of New Mexico Press, 1980); Gerhard Drekonja et al., *Ecuador Hoy* (Bogotá, 1978). On the recent return to civilian government: Thomas G. Sanders, "Ecuador: Politics of Transition," *AUFS Reports*, West Coast South America Series, Vol. XXIV, No. 1 (1977); Howard Handelman, "Ecuador: A New Political Direction?" *AUFS Reports*, No. 47 (1979).

7. CIDA, *Ecuador: Tenencia de la Tierra y Desarrollo Socio-Económico del Sector Agrícola* (Washington, D.C., 1965), pp. 412 ff. The role of latifundia became less important after United Fruit's sale of its plantations to the Ecuadorian government in the early 1950s. Big corporations are more dominant in the export sector (as opposed to bananas produced for domestic consumption) and control export marketing. Thus Companía Banenero Naboa (controlled by one of the most powerful families of the coastal export-banking elite) markets some one-third of Ecuador's foreign banana sales.

8. CIDA, op. cit.: John F. Uggen, "Peasant Mobilization in Ecuador: A Case Study of Guayas Province," Unpublished Ph.D. dissertation (University of Miami, 1975).

9. Carlos Larrea, "Estudio Social: Aproporción del Eccidente y Desarrollo Desigual en el Ecuador: Una

Primera Aproximacíon," Unpublished FLACSO paper (Quito, 1969).

10. The quote on Guayas fertility is from E. Erikson et al., *Area Handbook for Ecuador* (Washington, D.C.: Government Printing Office, 1966), p. 12. On the condition of rice tenancy see: M.R. Redclift, "Agrarian Class Structure and the State; the Case of Ecuador," Mimeograph (Quito: 1977); Also, M.R. Redclift, "Agrarian Reform and Peasant Organization in the Guayas Basin, Ecuador," *Inter-American Economic Affairs* 30 (Summer, 1976), p. 3-27.

11. CIDA, op. cit.; Uggen, op. cit.

12. *Censo Agropecuario Nacional*, Compiled in unpublished research by DEPALES (Quito, 1954).

13. On highland agrarian society, see: CIDA, op. cit.; Fernando Velasco, *Reforma Agraria y Movimiento Campesino Indígena de la Sierra* (Quito: El Conejo, 1979); Osvaldo Barsky, "Iniciativa Terrateniente en la Reestructuración de las Relaciones Sociales en la Sierra." In *Revista Ciencias Sociales*, Vol. 2, No. 5 (Quito, 1978), pp. 74-126; Barril et al., op. cit. I am greatly indebted to Professor Barsky (FLACSO and CEPLAES of Quito) for the many insights into the Ecuadorian sierra which he shared with me.

14. Agustín Cueva, *El Proceso de Dominación Política en el Ecuador* (Quito: El Solitierra, 1975), p. 59; César Raúl Robalino Gonzago, "El Desarrollo Económico del Ecuador" (Quito: JUNAPLA, 1969).

15. Prices paid to dairy farmers for their milk are state controlled and in recent years producers have complained that their profit margins have been greatly reduced. Production of cheese and other dairy products, however, continues to be very profitable and the largest dairy corporations (such as La Avelina) are vertically integrated from dairy farming through processing and packaging.

16. Piedad and Alfredo Costales, *Historia Social del Ecuador: Reforma Agraria* (Quito: Casa de la Cultura, 1971); Barsky, op. cit. The average size of the plots turned over in this manner was 3.2 hectares.

17. Muriel Crespi, "Changing Power Relations: The Rise of Peasant Unions on Traditional Ecuadorian Haciendas," *Anthropological Quarterly*, Vol. 44, No. 4 (1971).

18. Martz, op. cit., and Barsky, op. cit. Both note the low level of political mobilization among Ecuadorian highland peasants. Mobilization has clearly been more common on the coast, see Uggen, op. cit.

19. Following the Cuban Revolution agrarian reform laws were passed in the early 1960s in Chile, Colombia, Peru, and Venezuela. None of these led to significant changes in land tenancy. Only Bolivia (as a result of the 1952 Revolution), Cuba (the 1959 Revolution), Peru (under a left military regime in the 1970s), and now Nicaragua (the 1979 Revolution) have ended latifundia domination.

20. On the political debate over agrarian reform, see: Osvaldo Barsky, "Los Terratenientes Serranos y el Debate Político Previo al Dictado de la Ley de Reforma Agraria de 1964 en Ecuador," (Quito: FLACSO, 1978); Gustavo Cosse, "Reflexiones Acerca del Estado, El Proceso Político y la Política Agraria en el Caso Ecuatoriano, 1964-1977" (Quito: FLACSO, 1979).

21. *El Comercio* (Quito: November 17, 1961; December 6, 1961; December 16, 1961).

22. *El Telegrafo* (Guayaquil: June 12, 1963).

23. Ibid. (July 4, 1963).

24. Huasipungueros with over ten years of service on a hacienda were legally entitled to their plots and a cash payment from the landlord. See Barsky, "Los Terratenientes, " p. 48.

25. Barsky, "Iniciativa Terrateniente," p. 96; and CIDA, op. cit.

26. Velasco, op. cit., p. 99.

27. Velasco, ibid.; Alonso Rodrigo Vallejo, "La Lucha Campesina de Toctezinín," Unpublished thesis (Quito: Universidad Católica, 1978).

28. Barsky, "Iniciativa Terrateniente," p. 87.

29. Uggen, op. cit., pp. 210-226.

30. Redclift, op. cit. In the late 1950s there was a substantial increase in the role of mechanized rice estates. But these plantations still account for only 15 percent of cultivation.

31. As with all of Ecuador's agrarian reform legislation, most growers were entitled to compensation.

32. Velasco, an oratorically skilled but administratively weak populist, was elected president five times and ousted by the military before completing four of those terms.

33. On the 1972-1979 juntas of Rodríguez Lara and Poveda, see Sanders, op. cit.

34. Cosse, op. cit.; Redclift, op. cit.; Barsky, "Los Terratenientes."

35. Cosse, op. cit., p. 61; Banco Central. In 1979, the IERAC budget was cut to S/226 million (about $10 million) from a 1978 level of S/312 million; when inflation is taken into account this amounts to a near 40 percent cut in real terms.

36. *El Comercio* (June 8, 1979); and *Latin America Economic Report* VI (December 12, 1978), p. 398.

37. Data in this section are drawn from: Cosse, op. cit.; Ministerio de Agricultura, JUNAPLA, IERAC, *Evaluación de la Reforma Agraria Ecuatoriana, 1964-76: Resumen del Informe General* (Quito, 1978), henceforth referred to as *Resumen del Informe*; Banco Central.

38. Gonzalo Ortiz Crespo, "El Campesino y El Desarrollo Ecuatoriano" (Albuquerque, N.M.: University of New

Mexico Press, forthcoming); Benigno Cárdenas, "Desarrollo Rural: Concepto y Modelos Alternativos," in Barril et al., op. cit.; Velasco, op. cit.

39. The 1974 data from the United Nation's Economic Commission for Latin America (ECLA) showed Ecuador in about the same relative position in Latin America and actually suggested a lower average nutritional level—1,948 calories and 43 grams of protein daily—than in 1970. ECLA, *Latin American Development and the International Economic Situation*, Vol. I, Pt. I (New York, 1975). One study drawing on data from FAO, *FAO Production Yearbook, 1966* (Rome, 1967); AID, *Economic Data Books* (Washington, 1967); and USDA, Economic Research Service, *The World Food Budget, 1970* (Washington, 1964) rated Ecuador 103rd of 107 nations in the world in average daily caloric consumption in the early 1960s. This author finds that evaluation (placing Ecuador below India, Ethiopia, Indonesia, etc.) dubious. But, Ecuador is clearly among the three or four most malnourished nations in the Western hemisphere.

40. *Financial Times* (London: February 28, 1975) quoted in Ortiz Crespo, op. cit.

41. United States Department of Agriculture, Economic Research Service, *Agriculture in the Americas* (Washington, 1976), Tables 3-4. That 3.6 percent rate of growth actually placed Ecuador ahead of the Latin American average rate of 2.5 percent or the Andean nation rate of 2.8 percent. Much of that gain, however, was in livestock and export crops.

42. *El Comercio* (October 9, 1978; March 6, 1979). The sharp drop in milk output reversed an upward trend in recent years and was due to the severe drought which had badly hurt pastures. Meat production was sharply up in 1978 and 1979 (contributing to the overall agricultural growth rate in 1978 of 2.8%) as hard-pressed farmers slaughtered much of their herds.

43. See the statement by the Pichincha Chamber of Agriculture (Quito) in *El Comercio* (October 20, 1978).

44. Blasco Peñaherrera, "La Crisis Agropecuaria," *Vistazo* (Quito: November 24, 1978). Dr. Peñaherrera is a leader of the Liberal Party and was originally the party's vice-presidential nominee in the 1978-79 presidential race.

45. Comparing the value of production per hectare (Tables 9-10) is a somewhat imperfect measure of productivity since the units may be raising crops (or livestock) with very different commercial values. Unfortunately, I know of no data which correlate productivity of *particular* crops with the size of farming units. However, there is no evidence that smallholders systematically grow more commercially valuable crops (the opposite is likely the case). Ideally, controls should also be introduced for the quality of the land. Here again the evidence strongly suggests that minifundias tend to have inferior land, making their higher productivity rates more impressive.

46. Fausto Jordan, "La Agricultura en el Ecuador," in Drekonja et al., op. cit., p. 280. Obviously not all latifundia land is suitable for cultivation or pasture, but in Ecuador, as in most of Latin America, experts feel that large estates systematically underutilize their land. This is particularly true of the latifundia in the tropical region west of the Andes where large landowners control huge amounts of land "in reserve."

47. Barril, op. cit., pp. 68-70.

48. Francis Moore Lappé's *Diet for a Small Planet* has brought widespread popular attention to the fact that cattle raising—particularly for beef but to a far lesser extent for dairy as well—is a highly inefficient means of converting grain to protein or calories. While milk production in the highlands can be nutritionally justified, the conversion of much of the country's best potential crop land to pasture is of dubious social utility in a society plagued by severe caloric deficiencies.

49. Since the price of milk is state controlled (and kept at a reasonably low level; and cheese prices can rise freely, increasing amounts of milk are going into cheese and other dairy products that are far too expensive for most Ecuadorians. A more rational policy, according to many agricultural experts, would be to allow dairy farmers a slightly higher price for their milk (as a stimulus to production), expand the importation of cheaper powdered milk (to be mixed with local milk for commercial sale), reduce the retail price of milk through a state consumer subsidy (paid for out of petroleum revenues), and place a price ceiling on cheese and other dairy products that would make them unprofitable (or less profitable).

50. Cosse, op. cit. See Table 3 of this *Report*.

51. Assad Bucaram—chief officer of the national Congress and the long-time strongman of the CFP (now Ecuador's largest political party)—has his core political base in the slums of Guayaquil. Formerly the mayor of Guayaquil, Bucaram strongly criticized the 1970 agrarian reform Decree 1001 (ending rice sharecropping) and attacked government aid to rice-producing peasant cooperatives on the grounds that both measures increased the cost of rice to the urban poor. See Redclift, "Agrarian Reform and Peasant Organization." Urban voters are not against agrarian reform per se but fear it will increase their food prices.They are, of course, even more opposed to paying higher farmgate prices to agricultural producers. Given the fact that Ecuadorian smallholding peasants (like their counterparts throughout Latin America) have far lower incomes than even the urban working class, however, mandating higher prices for peasant-produced crops would be a justifiable form of income transfer from city to countryside in a region where urban per capita income typically outstrips rural incomes by five to one. It could also stimulate production. Since the Roldós administration upon taking office nearly doubled the legal minimum wage, higher farmgate prices seem even more justified and could be partially offset by consumer subsidies on lower class staples (bananas, rice, bread, and potatoes).

Food Policy Decision-Making in Colombia

Thomas G. Sanders

The process of decision-making in Colombian food and agricultural policy involves the interplay of a variety of participants and factors.[1] The most important participants come as individuals from the elite who dominate society and politics, but the institutions through which they function include the presidency, the technical sectors of government, organized interest groups, political parties, and the Congress. Other influences that contribute to the process include historical circumstances, peasant activities, and international lending agencies.

In order to bring into focus the decision-making process and to disentangle the roles of the diverse groups that influence policy, it is helpful to distinguish the four types of food and agricultural policy that have been significant since 1930:

1. Government efforts to encourage the use of land in a more intensive and productive way;

2. Agrarian reform;[2]

3. Government controls on credit, prices, imports, and exports that are periodically revised and have an effect on agricultural and livestock production;

4. Integrated Rural Development (DRI), a recent program aimed at increasing production among small peasants, and the associated National Nutrition Plan (PAN),[3] which is designed to improve food consumption in selected segments of the poorest population.

Each of these policies is associated with different historical contexts and reflects somewhat different concerns and pressures. When all are taken into consideration, they reveal the diversity of Colombian food policy, and they provide a framework for making some generalizations on how decisions are made. A completely satisfactory analysis is impossible, however, until more comprehensive empirical studies have been completed.[4]

The Colombian Political System

The Colombian political system within which food policy is shaped is usually depicted as a distinctive example of continuous leadership by an upper-class modernizing elite, based on long-dominant families, that has succeeded in retaining power despite periodic electoral changes and various challenges to its hegemony.[5] In other major Latin American countries, by contrast, the traditional upper class has had to yield at least part of its control and allow the participation of new classes and interest groups. Though Colombia's upper class has experienced many similar threats to its domination—populism, Marxist movements, organized labor, and reformist military interventions—the governing elite has managed to prevent them from seizing control or even sharing significantly in power.

From 1958 to 1974, the political mechanism for elite cooperation was the National Front, an agreement between the Liberals and the Conservatives, the two dominant parties, to share power and divide the principal offices equally between them. Though the two parties had conflicted historically over the role of Catholicism in national life, and though the differences between them had led to a breakdown of democratic processes between 1953 and 1958, the leaders of both parties, mostly from upper-class backgrounds, came to recognize that their mutual interest in stability and maintenance of the economic system was better served by collaboration. This "consociational" pattern continued after 1974, when the formal division of offices under the National Front ended, as the two Liberal presidents still chose to include the Conservatives in political offices and decisions.

Nevertheless, to affirm that an upper class elite controls Colombia does not fully explain the motivations and tensions that underlie the process of making policy. The Colombian elites have often been described as realistic, in the sense of altering their obvious interests in response to circumstances that seem to threaten their control of the system. They have also been considered by many analysts as enlightened and paternalistic in pursuing policies that seemed to have served the national interest and thus to some extent the interests of other socioeconomic groups. Despite ongoing problems of economic inequality, recurrent social tension, and political restrictions, the elite has guided Colombia for the past 20 years with relative institutional stability and continued economic growth.

Within the elite itself, cleavages exist that provide impetus to political competition which in turn influences policy. Industrialists and agriculturalists often compete in the political process when their interests clash (though sometimes prominent individuals have investments in both industry and agriculture). Within the agricultural elite itself, traditionalists and modernizers have long been in tension; since the 1930s, however, relatively modernized commercial producers have almost totally replaced the traditional feudal landowners. The various agricultural sectors express their concerns through specific associations (*gremios*)—coffee producers, sugar plantation owners, cattlemen, and so on—whose interests sometimes also conflict.

Various political institutions too are engaged in a struggle for relative power and influence that at times has an effect on food policy. These include the Liberal and Conservative Parties and the principal factions within them. The Liberals, for example, are currently divided into three personalistic currents with somewhat different orientations of policy and style—Lleristas, Lopistas, and Turbayistas (adherents of the last three Liberal presidents, Carlos Lleras Restrepo, Alfonso López Michelsen, and Julio César Turbay Ayala). The Conservatives have two chief factions: the more "progressive" group called Ospinistas or Pastranistas (after former Presidents Mariano Ospina Pérez and Misael Pastrana Borrero) and a more "traditional" wing, the Laureanistas or Alvaristas (from Laureano Gómez and his son, Alvaro Gómez Hurtado).

In recent years, divisions within the structures of government itself have also affected the way policy is made. Congress has lost power as a result of many factors: the National Front arrangement; the inability of the various factions to reach agreement in Congress; the need for "technical" rather than "political" decisions on economic aspects of development; and the assumption of greater power by presidents who by-passed Congress to govern by decree under "state of siege" and "economic emergency." In the bureaucracy, a more technologically trained and oriented group based in the economics ministries and such autonomous agencies as the National Planning Department (DNP) has come to exercise a more significant role in formulating policy than the "political" bureaucracy, the traditional arena of competition for political posts and for the functioning of the patron-client system.

The ascendancy of the president and the technocrats is associated with their increased awareness that broad national economic necessities, as they perceive them, do not necessarily correspond to the demands of particular interest or political groups. Policy conflicts often stem from clashes between particular group pressures and the commitment of top administrators to what they see as necessary for other national goals, especially economic ones.

Finally, acknowledging the dominant influence of the upper class in Colombian politics should not obscure the inputs of the middle class and the peasants, or the reasons why these other classes lack significant impact in altering policies affecting food production and the rural areas generally.

In Colombia, the middle classes, as part of their upward aspirations, tend to imitate the values and outlook of the upper class rather than to challenge them. Many of the most capable individuals from the middle class have been coopted by the politico-economic elites. That many are also alienated from the system is reflected in the low level of voting in national elections (about a third of the electorate participates). When middle class groups have asserted themselves publicly, however, as in teacher, student, or white collar strikes, their effectiveness has been countered by the continued multiclass strength of the two traditional parties, the loyalty of the armed forces to the system, and the government's use of "state of siege" and "economic emergency" powers when it encounters opposition.

The role of peasant unions in the development of food policy will be discussed later in more detail, but for the moment it is sufficient to note their relative lack of significance. Although peasant problems and unrest frequently provide a signal to the dominant elites that leads them to change policies, peasants as individuals and their organizations do not participate directly in formulating policy or in decision-making. Peasant organizations, moreover, are divided by religious and political differences that also limit their effectiveness.

* * * * *

Policies related to food production have not been peripheral aspects of modern Colombian life; indeed, they have been central political and economic issues. Although many agricultural questions are associated with strong tensions among the groups that have been described, in other cases a relative consensus has been reached that has reflected the mutual interests of the dominant groups and has contributed to overall increases in agricultural and livestock production.

The Development of a Commitment to More Productive Land Use

The most significant example of a consensus affecting food policy is the now unquestioned assumption that intensified land use, modern farming methods, and increased agricultural production within a capitalist framework should be encouraged.

In his well-known analysis of Colombian agricultural policy through the 1950s, Albert O. Hirschman points out that even though large landholdings were historically dominant in Colombia, a strong bias in favor of using them productively goes back to the colonial period. The Spanish authorities consistently upheld the view (though they could not enforce it) that title to land was contingent on the owner's using it, and this attitude was carried over when the colony became an independent republic. It was confirmed by one of the most important processes in national history and self-definition, the internal migration of the nineteenth century in which enterprising pioneers occupied unused land in Antioquia and Caldas and converted it to the production of coffee, which rapidly became Colombia's most important economic resource.

In the period after World War I, new ideas and circumstances reinforced the consensus in favor of effective use of unproductive lands. In the 1920s, Colombians were influenced by ideological currents emphasizing the government's dominant role in promoting national development through industrialization, the modernization of traditional structures, and new forms of labor relations. Among the political elites such attitudes were especially strong in the younger generation and in the Liberal Party, which was more urban in orientation than the Conservatives and thus more exposed to outside ideas.

These sentiments were encouraged by economic and social circumstances. A period of rapid economic growth in the 1920s, which was based on domestic public works and coffee exports, was followed by the depression and a sharp drop in coffee prices and foreign exchange earnings. In the countryside, where a "traditional" system based on large landholdings, low production, and semifeudal forms of labor organization was prevalent, unrest and violence became widespread: peasants squatted on unused land, refused to pay rent, and insisted on better working conditions. Food shortages and high prices compelled the country to import food. Colombia's leaders concluded that, despite its extensive land resources, the nation was not using the land as it should.

Even the Conservative governments that dominated Colombian politics from 1902 to 1930 gradually came to realize that the traditional order was in crisis, that production had to be encouraged, and that an outlet had to be provided to avoid violence and meet peasant aspirations for land. In various decisions the highest courts rejected many long-standing claims for land tenure by landlords who did not fulfill the legal requisites, deciding in favor of individuals who were actually cultivating the land. The growing impact of coffee production on the national economy, increased urbanization with its concomitant demand for food, and the need for agricultural raw materials for industrialization helped build a broad constituency that gave priority to productivity over traditional claims to the land.

In 1934, the reformist Liberal Alfonso López Pumarejo became president, bringing into his administration a group of young intellectual activists. Congress, though dominated by large landowners, passed Law 200 of 1936, which denied that property was an absolute right and allowed for expropriation if the public interest required it. One of its major provisions established the principle that those who occupy land

and use it are presumed to be the owners. Squatters could not be evicted unless the owner proved title before the year 1821, which few could. Another provision required that lands of over 300 hectares which remained uncultivated for 10 years would revert to the public domain.

The example of Law 200 illustrates certain key factors affecting agricultural policy that have continued in somewhat different form to have influence in subsequent decades. The landowners, who were the dominant segment of the governing elite, confronted a turbulent situation in the countryside which led them to believe that, without change, radicals would come to dominate, and that they might lose control totally of their property and of the system. The high priority given to productive use of the land by laws and court decisions provided a precedent for policymaking. Food shortages and inflation were concrete indicators that idle land had to be put into production. Nationalism impelled the use of agriculture to improve living standards and stimulate industry. A strong president, assisted by advisers with modern ideas, swayed Congress and pushed through the legislation.

Law 200 has frequently been criticized for leading to widespread eviction of squatters and for failing to bring about the revolution in land tenure that certain of its provisions implied. However, the best analysts of this period, like Hirschman and Salomon Kalmanovitz, indicate that many owners of extensive idle land sold part of it to peasants or to more production-oriented entrepreneurs. They were inclined to work out arrangements with the squatters, sharecroppers, and renters who occupied their land out of fear that in 1946 (10 years after the legislation) they might lose it.[6]

Because of declining import capacity, World War II stimulated industrial expansion, especially in textile and food processing sectors, which required agricultural raw materials. High food prices during and after the war also led landowners and peasants to increase their production. The government supported the trend by making rural credit available, importing tractors, and investing in infrastructural improvements such as roads and irrigation in agricultural zones.

After 1946 Colombian politics polarized as the Conservatives took advantage of a Liberal split to recapture the presidency, and opposition to Conservative policies coalesced around the populist leader Jorge Eliezer Gaitán. Gaitán's assassination in 1948 led to an unprecedented bloodbath, "La Violencia," the Violence. Though it had a profoundly disruptive effect on rural life, it also had certain positive influences on production. One was a continuation of the tendency by owners of large landholdings, especially absentees, to sell their property or abandon claim to it. In some instances squatters or smallholders gained possession, and in others commercial entrepreneurs took advantage of a drop in land prices to expand their holdings. A second effect, like that of the 1920s, was a renewed sensitivity by the dominant elites to the reality of rural violence and the necessity to adopt policies to assure political stability. This, coupled with the specter of the 1959 Cuban Revolution, eventually led to the agrarian reform legislation of 1961. Finally, La Violencia intensified the exodus to the cities, which led to further industrial growth and steadily increasing demand for food and industrial raw materials.

A major shift in the geography of Colombian agriculture also occurred during this period. Commercial crops, many scarcely produced before in Colombia, expanded in lowland zones like the Cauca and Magdalena Valleys and the Atlantic Coast, displacing to less fertile areas the cattle that had previously occupied these regions. As a result, from 1950 to 1972, agriculture, apart from coffee, grew at 4.6 percent annually, with 2.0 percent based on increased hectarage and the rest on productivity.[7] Colombia became an important producer and in some instances exporter of cotton, sugar, rice, soybeans, sorghum, barley, and oil seeds. Since many of the commercial crops were new, the producers were not inhibited by traditional methods and adopted machinery and modern technology from abroad. As Table 1 reveals, commercial farmers (but not plantations) led the way in production gains. The availability of foreign exchange and the restoration of international commerce after World War II intensified the use of tractors and other types of farm machinery.

The growth of commercial agriculture also had an effect on the rural labor force by increasing the demand for salaried workers. Between 1951 and 1964, the number of agricultural workers totally dependent on salaries increased by 140,000, even though this was also a period of extremely heavy urban migration. By 1973, salaried workers (*obreros*) totaled about one million or 49 percent of the economically active rural

Table 1
Indexes of Production According to Types of Crops

Year	Commercial crops	Traditional crops	Plantations	Mixed Forms	Total (without) coffee)	Coffee	Total (with coffee)
1950	58.8	88.9	71.6	64.6	73.7	72.1	73.1
1951	72.1	99.3	73.6	84.0	87.1	64.4	79.8
1952	77.2	102.0	83.0	98.0	95.2	86.0	92.3
1953	74.7	103.0	89.3	91.6	92.8	81.9	92.5
1954	86.9	96.0	91.7	87.5	90.8	86.0	88.7
1955	83.2	96.0	95.7	97.0	93.2	80.4	88.7
1956	90.1	98.8	99.0	92.8	95.0	71.4	85.4
1957	87.7	97.0	99.1	91.6	93.5	77.8	87.1
1958	100.0	100.0	100.0	100.0	100.0	100.0	100.0
1959	167.6	102.6	106.0	107.5	123.1	98.5	115.4
1960	168.7	101.5	111.8	110.3	125.7	102.3	117.8
1961	187.2	109.0	113.7	90.5	127.6	96.0	118.1
1962	206.6	111.4	109.1	116.2	143.2	102.8	131.7
1963	191.6	111.2	120.0	101.8	131.3	96.0	121.7
1964	186.8	92.0	116.0	98.3	122.4	99.8	116.4
1965	201.5	103.0	131.3	115.0	141.9	105.0	132.2
1966	229.8	110.7	143.6	111.1	152.3	97.2	138.6
1967	242.3	114.4	148.5	116.7	158.3	101.7	143.7
1968	288.8	115.0	151.0	121.6	180.4	102.3	160.7
1969	302.4	127.0	123.5	122.6	183.2	102.3	162.5
1970	310.1	132.6	131.6	129.7	188.6	106.8	163.6
1971	299.6	120.1	137.3	127.4	187.0	103.4	164.1
1972	367.9	129.2	141.3	119.3	212.0	105.5	183.2
1973	330.7	134.5	144.8	116.2	211.8	111.3	185.3
1974	402.7	131.3	153.5	111.6	239.9	117.7	207.5
1975	366.7	123.8	157.4	117.5	224.1	113.8	195.0
1976	395.3	121.0	177.6	129.9	238.5	108.7	204.2
	r = 7.6			r = 4.6			r = 4.03

Source: Kalmanovitz, p. 308.

Table 2
Economically Active Rural Population
1938, 1951, 1964, and 1973
(in thousands)

Category	1938 Number	1938 %	1951 Number	1951 %	1964 Number	1964 %	1973 Number	1973 %
TOTAL	1,808.9	100.0	2,023.2	100.0	2,405.2	100.0	1,870.6	100.0
1. Employers	307.4	17.0	335.1	16.1	307.7	12.8	155.1	8.3
2. White-collar workers	9.6	0.5	28.8	1.4	48.1	2.0	73.0	3.9
3. Manual workers	502.1	27.7	818.2	40.4	967.8	40.2	913.5	48.8
4. Self-employed workers	609.5	33.7	488.0	24.1	706.6	29.1	299.4	16.0
5. Domestic workers	338.7	18.7	278.1	13.7	367.8	15.3	191.2	10.2
6. Miscellaneous	41.6	2.3	75.0	3.7	7.2	0.3	238.3	12.7

Source: ibid., p. 38.

population and was clearly the predominant rural class. An estimated third of the owners of plots also earned part of their income from salaried work.[8]

Although "traditional" owners of idle land represented a powerful pressure group in the early 1920s, the incapacity to defend their position against the law, peasant violence, political leadership that lacked sympathy for them, and the expansion of a class of capitalist producers sensitized to the market undermined their position. Elite opinion had arrived at a consensus that their own self-interest and national development were served by better use of agricultural land. Normally, if agriculturists could show they were making a serious effort to produce, even though they might not be very efficient, the government protected them. Land invasions would be repressed by the police and by the central government if the owner could demonstrate utilization of his land, but idle land was even more likely to suffer invasions and the public authorities were less willing to support the owner's claim to title.

The Tax Question of the 1950s

During the 1950s, agricultural policy was dominated by political debate over the use of taxes to stimulate production. This issue illustrates the impetus in Colombia toward agricultural modernization, but it also provides insight into a new factor in food policy-making, the influence of international lending agencies on domestic decisions.

After World War II, foreign economic missions began to exert significant influence in Colombia, the most important early one being a delegation from the World Bank in 1949 led by North American Professor Laughlin Currie. Although the existent pattern of land use was already changing rapidly, the Currie mission considered Colombian tenure practices irrational and proposed that the tax on land be revised on a sliding scale that would reward production and penalize nonutilization.

In his discussion of this period, Hirschman points out that the proposal stimulated wide debate among Colombia's political elite, who were impressed with the views of outside experts and assumed they would benefit from large-scale loans if the recommendations were adopted.[9] The World Bank mission shifted national debate to fiscal rather than confiscatory policies in land tenure, but the Colombians realized they had to alter the suggestions to meet the situation. The chief practical problem lay in the lack of a reliable cadastral survey to evaluate landholdings and the almost insuperable difficulties in conducting one.

Numerous tax schemes were proposed and one, Decree 290 of 1957, was promulgated by the military junta that overthrew the populist military regime of General Gustavo Rojas Pinilla. However, Decree 290 was based on a complex classification of land for which there was insufficient information, and when democratic processes returned in 1958 with the National Front, the new minister of agriculture refused to enforce it. During this period, disputes over the appropriate type of land tax, the opposition of most landowners to any tax changes at all, the absence of clear beneficiaries to pressure for it, and factional political conflict all contributed to the failure of the land tax proposals.

Hirschman argues, nevertheless, that the common expectation of a heavy tax on poorly used land, plus other more feasible measures by the government, provided a further stimulus to agricultural modernization. Decree 290 contained provisions that went beyond the recommendations of the World Bank by allowing income tax deductions for a variety of agricultural investments, and reductions in income and property tax obligations for certain agricultural activities. The Decree also aimed at encouraging the establishment of partnerships or corporations that would lease lands from large traditional landowners and grow commercial crops on them.

By the 1960s, Colombian agricultural production was being modernized, and a consensus was firmly established among decision-makers that the trend was a good thing. Precedents of government intervention through credit and feasible tax stimuli were established. The real impetus for commercial agricultural development came, however, from landholders and entrepreneurs who responded to the demand for food, raw materials, and exports.

An initial attempt to assist smaller producers within the framework of the modernizing consensus led to Law 20 of 1959. Its purpose was to channel credit to small- and medium-sized producers by requiring the Agrarian Bank (Caja de Crédito Agrario, Industrial y Minero, usually called the Caja Agraria) and the commercial banks to dedicate 10 percent of their loans to these sectors. Referring to agricultural policy precedents, Law 20 reiterated that all agricultural land should be "efficiently used" and that the state could appropriate without compensation land that was not adequately exploited.

The results on relatively large landholdings oriented to export products were nevertheless more impressive than those on smallholdings specializing in consumer items. As the 1956-1976 data in Table I indicate, "commercial" items—rice, cooking oil, barley, sorghum, soybeans, and sugar (many of these exports)—increased their production at a substantially higher rate than "traditional" crops—beans, manioc, bananas, *panela* (popularly consumed sugar), or "mixed" crops (corn, potatoes, wheat).[10] This suggests that intensified land use, modernization of production, and the government stimuli associated with them tended to be used by and to benefit the larger farmers. The fact that small producers for domestic consumption were not increasing production at a comparable level prevented the nutritional standards of the population from improving substantially.

Productive use of land and agricultural modernization, then, are ideological constants that have channeled policy discussion for the past two decades. There are no clear opponents to these principles because the traditional *latifundio* (large estate) has almost disappeared. Furthermore, productivity has many practical advantages in Colombia, with its rapidly expanding population and cities, among them the increase in national income from exports, both legitimate (coffee, sugar, soybeans) and illegitimate (marijuana, cocaine, and other forms of contraband), in demand from Colombia's wealthier neighbor, Venezuela. Modernization also fits the interests of the dominant elites who have rural enterprises or control industries and commerce that benefit from agricultural development. It similarly conforms to the orientation of government technocrats who need a justification for their policies and are sensitive to the same interest in international lending entities. The reformers and peasants who pushed for agrarian reform in the 1960s also included modern inputs in their program because, without production stimuli, land distribution would not be effective. The urban middle and lower classes benefit in the form of cheaper, more abundant foods. Colombian agricultural policy, consequently, even when influenced by other pressures and influences, begins with a commitment in principle to increase production through the effective use of land and modern technology.

Agrarian Reform

Given the difficulties in agreeing on a land tax in the 1950s, it is ironic that agricultural policy in the 1960s was dominated by an even more radical proposal, agrarian reform.

The country's political leaders realized by the late 1950s that taxing schemes, while stimulating production on large holdings, would have little or no effect on the smaller peasants. The latter required special attention, especially those who had no land at all or only very small plots that could not provide an adequate living. By 1960, furthermore, the context had changed. The Cuban Revolution had been consolidated, and in its wake there was widespread expectation that new insurrections stemming from rural discontent would take place in other Latin American countries. The Organization of American States, in meetings in Bogotá in 1960 and Punta del Este in 1961, announced its support for agrarian reform. The United States government included agrarian reform as an essential plank of the Alliance for Progress, in the absence of which it allegedly would not provide financial assistance. Colombia's experience with the Violence, in which perhaps 200,000 people were killed, and the military interregnum of 1953 to 1958 verified to the governing elites that maintaining the system required eliminating the causes of rural dissatisfaction.

That the peasants were unhappy was reflected in the elections of 1958. Alfonso López Michelsen, son of former President López Pumarejo, had organized a left-of-center faction of the Liberal Party, called the Liberal Revolutionary Movement (MRL), which made a good showing in the countryside. Leaders of long-established communist rural enclaves near Bogotá ran for office on the MRL ticket and won.

In both the Liberal and Conservative Parties, factions that favored some type of agrarian reform increased their strength in Congress. The most influential personality among the Liberals was the dynamic Senator Carlos Lleras Restrepo, who had been one of the young advocates of agrarian change in the early 1930s and the leading congressional spokesman for Law 200 of 1936. In the other party, the "Progressive" Conservatives led by Mariano Ospina Pérez, also gained ground and became dominant over the more traditional faction.

On August 31, 1960, Liberal President Alberto Lleras Camargo appointed a National Agrarian Committee. It was stacked with agrarian reform supporters, and its chairman was Carlos Lleras Restrepo. After extensive debate and revision in the Committee and Congress, the legislation was passed and signed as Law 135 on December 13, 1961.

Despite the passage of Law 135, agrarian reform has not been a success in Colombia; indeed, in many respects no reform has occurred. The explanation is complex, but the fundamental cause has been resistance by the large landholders to its application. On the assumption that this social group has an important influence on agricultural policy through its representatives in Congress, its sectoral organizations (*gremios*), and its slots on the directing boards of the implementing institutions, it is necessary to look at the record of agrarian reform. Just as international pressure, political leadership, and various circumstances led to the passage of the legislation, changes in these factors also inhibited its application and led to its eventual abandonment.

In 1960 and 1961, a coalition of Liberals and Ospinistas pushed agrarian reform through Congress, but at the time there was strong opposition from the Laureanist Conservatives, who represented the most traditional landed elements, and from the MRL of López Michelsen. Despite his leftward stance and his rural support, López Michelsen had never favored agrarian reform, preferring instead colonization for the landless and credit with technical assistance for the small peasant. Furthermore, many of those who voted for the measure were lukewarm or carried along by the tide of circumstances. One of them, an independent senator from Antioquia, said that he voted for the bill because it was better than nothing, but added with great perception:

> *This country is not prepared either socially or technically for an agrarian reform.... A complete indifference concerning the entire matter was exhibited by the* campesinos. *Actually, this statute is more urgent for the directing class than for the directed.*[11]

A second factor that influenced the acceptability of the legislation to the Congress but undermined its effectiveness was the system of priorities in land acquisition and the method of payment that were adopted. Agrarian reform programs affected first easily accessible publicly owned land, then privately owned but uncultivated land, poorly cultivated land, and only as a last resort properly cultivated land. The fact that some properly cultivated land probably would have to be expropriated, because it had already been invaded by tenants or was located in regions of great population density, made the question of payment the most disputed issue of the committee hearings and congressional debates. The central problem concerned the provision for partial payment in bonds, which would lose value, even with interest, under the 10 percent annual inflation rate of this period. In a concession to the landed interests, the matter was solved by guaranteeing 20 percent of the payment to owners of properly cultivated land in cash and the balance in special treasury obligations that did not depreciate. These generous provisions reveal that in Congress (where most of the members were landholders or sympathetic to their interests), many of those who accepted the proposals and worked out the concessions assumed that agrarian reform would affect unused land and have little impact on cultivated holdings.

Implementation of the law, furthermore, got off to a bad start when shortly after its passage Conservative leader Guillermo Leon Valencia became president under the National Front agreement for alternation of offices between the two parties. Although he had voted for agrarian reform, Valencia was not a strong advocate. He is generally considered to have been an inactive, indecisive president inclined to rhetoric, whose administration was plagued by economic problems, strikes, and the first signs of disenchantment with the National Front. Within the Colombian Agrarian Reform Institute (INCORA), which was supposed to apply the law, the allocation of posts on the board of directors according to political factions hobbled its land redistributive activities.

In 1966, Carlos Lleras Restrepo won the presidential election and entered office determined to stimulate and amplify agrarian reform.[12] After 1968, the pace of land distribution picked up, as INCORA carried out over half of all negotiated purchases and almost two-thirds of the expropriations in its first 10 years. By June 1970, 185,742 hectares had been purchased and 71,986 hectares expropriated. Yet, this totaled less than one percent of the country's agricultural and livestock area. The number of peasant beneficiaries was very limited as well. Though statistics on the number of families who benefited from these acquisitions vary, one estimate by a

Table 3
Distribution of the Land: Number of Plots, Area, According to Size

	Holdings				Area			
Size of holding (Hectares)	1960 Number	%	1960 Hectares (1,000)	%	1970-71 Number	%	1970-71 Hectares (1,000)	%
TOTAL	1,209,672	100.0	27,337.8	100.0	1,176.811	100.0	30,993.2	100.0
10 & under	925,750	76.5	2,403.7	8.8	859,884	73.1	2,234.3	7.2
Between 10 & 20	114,231	9.4	1,572.1	5.7	117,863	10.0	1,598.6	5.2
Between 20 & 50	86,789	7.2	2,638.7	9.7	100,010	8.5	3,054.5	9.9
Over 50	82,902	6.9	20,723.4	75.8	99,054	8.4	24,105.8	77.7

Source: Kalmanovitz, p. 54.

writer in 1975 in Colombia's most prestigious economic journal was 12,570.[13]

INCORA took over most of its land from tracts already owned by the state and by *extinción de dominio* (failure by the alleged owner to put his land into effective production). By the latter means, INCORA eventually acquired a total of nearly 3,500,000 hectares, but most of it was located in parts of the country that were lacking in infrastructure.

A far more significant form of land redistribution in Colombia resulted from INCORA's regularization of the titles of persons who had previously invaded uncultivated holdings. These decisions had affected 100,678 families by 1970.[14]

Table 3 demonstrates that, despite President Lleras Restrepo's good intentions, the Colombian agrarian reform program did nothing to redistribute land in its first decade. In 1960, large holdings (over 50 hectares) constituted only 6.9 percent of all agricultural units, yet controlled 75.8 percent of the land. At the same time, smallholders (under 10 hectares) represented 76.5 percent of all farm units, but owned only 8.8 percent of the land. A decade later, as Lleras left office, large units composed 8.4 percent of the farms and had actually increased their proportion of the land to 77.7 percent. At the same time, smallholders (now 73.1 % of all farms) saw their share of agricultural lands *fall* from 8.8 percent to 7.2 percent. As we shall see, there has been no alteration of that pattern in the 1970s.

The principal financial allocations of INCORA went to irrigation, credit, and technical assistance, which were also considered aspects of the agrarian program. However, the concentration of INCORA's attention in these areas was in part a reflection of its incapacity to acquire land for redistribution where the pressure for it was the strongest.

Even these meager efforts were stifled when Lleras Restrepo left office in 1970. Circumstances had changed. Fear of Cuba's revolution had declined, and the Alliance for Progress was dead. Responding to the interests of the large landholders, the new Conservative administration of Misael Pastrana Borrero suspended all land redistribution programs.[15]

Though Pastrana represented the progressive wing of the Conservative Party, his "Four Strategies" policy offered an alternative approach to agricultural questions. The "Four Strategies" was adopted almost completely from a set of recommendations called "Operation Colombia" that had been formulated by former World Bank adviser Laughlin Currie.

Currie (and Pastrana) hoped to solve Colombia's agricultural problems by further improving the technological level of the large- and medium-sized producers while attracting the small peasants into the cities by offering them employment in massive housing construction projects. They considered the small peasant an anachronism in a modern society. Programs of land distribution, infrastructure, and credit, Currie argued, would only be "a cruel policy, unjust, and highly dangerous."[16] The problem of the rural lower class lay in its reduced income level. Pastrano and Currie insisted that by diverting public investments to housing, which was highly labor absorptive, peasants would migrate to the cities to take the jobs available, commercial agriculture would be stimulated to produce the food items they needed, and industry would expand to meet the increased demand for consumer goods. The housing program, they argued, would initially increase the purchasing power of the urban and rural lower classes, but in time they would be absorbed into the new jobs provided by the industry and modernized agriculture that their demand had helped stimulate. In this way, the small peasant class would gradually be eliminated. Large, commercial agriculture would expand and modernize not only because of domestic demand but also to fulfill another aspect of the Currie-Pastrana policy, increasing foreign exchange earnings through diversified exports, including agricultural products and cattle.

Although Pastrana presented himself as a reformer and claimed that the agrarian reform legislation would be implemented, his administration came to terms with the large landholders in a way that spelled the end of agrarian reform. Pastrana's proposals in this area focused on two matters: legislation clarifying the meaning of "adequately cultivated land," and a new tax on "presumed income" as a stimulus for landholders to produce.

After an initial period of inconclusive debate on these issues in the Congress, an agreement was reached at a private meeting held in Chicoral, Tolima, in January 1972.[17] Chicoral included representatives of the two principal parties, many of whom were large landholders, but no one was invited from peasant organizations (which at the time were strongly agitating for land redistribution). Nor was anyone invited from the Llerista, pro-agrarian reform wing of the Liberal Party. Lleras Restrepo, whose enthusiasm and ability had been decisive in the changes of 1936, 1960-61, and during his 1966-1970 term of office, was out of the country heading an office of agrarian reform in FAO. The Chicoral Pact helped clarify the criteria for utilized land and was even initially attacked by the Colombian Society of Agriculturists (SAC), who feared that it might lead to a more active reform policy. However, Chicoral and Laws 4 and 5 of 1973, which were passed by Congress in response to the recommendations of the pact, actually ended the expropriation of any land that could be proven to be cultivated by the owner.

Since 1970, no Colombian president has favored agrarian reform. During this time, the political influence of Lleras Restrepo in his own party has declined. The Lopista and Turbayista factions cooperated to gain control of the Liberal Party and guarantee that their respective leaders would be elected to the presidency in 1974 and 1978.

Peasant Organizations and Food Policy

Although Colombia has several national organizations of peasants, they have been unable to exert influence in favor of agrarian reform or to overcome the strength of the large landowners.

Peasants organized locally and regionally to pressure for their interests in the 1920s and 1930s, but the first national peasant organization was the National Agrarian Federation (FANAL), founded (along with a parallel urban workers' confederation, the Union of Colombian Workers [UTC]) in 1946 by Jesuit priests. The Catholic Church at the time was interested in developing its own labor organizations to counteract alleged leftist and Marxist infiltrations in the Liberal Party-sponsored Confederation of Colombian Workers (CTC).

FANAL adopted its orientation from Catholic social teaching and used priests as "moral" advisers. Though commonly criticized for being conservative, FANAL has actually followed a reformist trajectory—favoring redistribution of landholdings and defending many peasant land invasions. It has also stressed specifically Catholic preferences for cooperatives and worker-controlled enterprises. FANAL has a base in all the rural classes, such as salaried workers, sharecroppers, renters, and smallholders, and in 1979 claimed 60,000 members. Perhaps because of its Catholic orientation and general weakness, however, FANAL has never succeeded in developing a mass membership in the Colombian countryside. Though it has tried to avoid association with any political party, it generally supported the National Front and has taken positions on specific political issues, acting by petition, pressure, and demonstrations.

A second form of rural organization is the system of state-initiated communal action juntas.[18] These developed shortly after the National Front began in 1958 and were a government attempt to counteract violence by organizing communities to participate in various development activities like school construction, health programs, road construction, sports, and agricultural improvement. According to Bruce Bagley and Matthew Edel, the principal zones of government attention under this program were those where violence had been most prevalent. The juntas received a wide range of support from groups as diverse as the Coffee Growers Federation, the Catholic Church, and the Alliance for Progress.

The juntas continue to be the most common form of organization in Colombian rural communities, but they have never become a significant vehicle of united action by peasants. The government has carefully restricted them to local matters and curbed their attempts to develop into broader class organizations. The well-known revolutionary priest, Father Camilo Torres, tried to work through the juntas in the early 1960s, but gave up and joined a guerrilla liberation force. The government followed a policy of rewarding juntas that collaborated with it and punishing those that did not. Today, juntas of communal action can be found in nearly all rural communities, but they are intimately linked to the patron-client system of the political parties and oriented to getting benefits for their communities. Their officers often are individuals interested in political advancement rather than representation of their constituents' interests.

The principal stimulus for a comprehensive structure of rural organization came from President Carlos Lleras Restrepo (1966-1970) who, as part of his emphasis on agrarian reform and increased production among smallholders, promoted the National Association of Peasant Users (of Government Services)-ANUC.[19] Lleras had several motivations for creating ANUC. One was his sincere conviction as a social reformer that peasants should be organized. Another was his desire for institutions to represent peasant interests as a support for agrarian reform, in negotiations with employers, and on the directing boards of policy-implementing institutions. Finally, Lleras, who could not constitutionally succeed himself in the presidency but planned to run again in 1974, wanted to expand his political base as the defender of agrarian reform and peasant interests.

In the period from 1968 to 1970, ANUC inscribed nearly a million members as the government contacted the rural population that had used various kinds of services in the past and encouraged it to organize on various levels. Although ANUC was based on a comprehensive structure that extended from the local (*vereda*) to the national level, even Lleras considered it an institution to defend the economic interests of its members; it was not to adopt a political line independent from the government. This hope seemed to be realized when the early declarations of ANUC reflected a moderate tone consistent with the policy outlook of the Lleristas.

In the early 1970s, however, ANUC became politicized and radicalized, as the aims of the peasants and the policies of the Pastrana administration came into conflict. After 1968, the condition of many peasants worsened. Law 1 of 1968, passed under Lleras Restrepo, provided for enrolling sharecroppers and renters preparatory to expropriating land for them, but certain landowners, to avoid this outcome, responded by expelling them from the land. As a result, peasant land invasions increased. The Pastrana administration appointed union promoters to keep ANUC under control and undermine the Lleristas' influence. Furthermore, after the Pact of Chicoral in 1972 and the winding down of agrarian reform, ANUC responded by emphasizing land acquisition and other more independent demands through invasions, demonstrations, and similar methods.

In mid-1971, ANUC signaled its radicalization by agreeing on a document called the First Peasant Mandate, which revealed that the organization had moved well beyond the Llerista orbit. The Mandate advocated the formation of peasant-controlled Executive Councils of Agrarian Reform (CERAs) that would carry out their own expropriations (by invasions) without going through legal procedures and compensating owners. The peasants themselves would take charge of organizing and managing the institutions of the agrarian reform. The appearance of the First Peasant Mandate led to a definite break with the Pastrana government, which refused to publish it and withdrew the subsidy previously granted for printing the ANUC magazine, *Carta Campesina*.

The clash with the government reached its climax within a year as ANUC divided into two factions, one patronized by the government and called the Armenia line and the other, the Sincelejo line, in opposition to the government. (Armenia and Sincelejo are the cities in which each "line" was organized.)

ANUC-Armenia has evolved into a pocket organization of the government. Its current leaders claim to have 1,750,000 members, but this figure represents nothing more than the total of everyone who has in any way used government services in the past 20 years. Officials of ANUC-Armenia have their headquarters in the Ministry of Agriculture building and seem to be there because the government gives them offices and salaries. The Armenia faction is the beneficiary of the Colombian policy of allocating positions on the boards of directors of various agricultural entities to peasant representatives, but its spokesmen admit they have no influence against the weight of the other represented interests. ANUC-Armenia differs with the government in favoring agrarian reform, but its leaders do not press for it nor do they go into opposition because they claim that such a strategy would be counterproductive.

ANUC-Sincelejo was the most active peasant organization from 1972 to 1975, but it has declined drastically as a result of external pressures and internal divisions. In the early 1970s, ANUC-Sincelejo was principally responsible for organizing hundreds of land invasions that aroused the hostility of the landowners and the government against them. At the same time the ruling elite and the government were reaching a consensus that land expropriation would not be part of the agenda. By radicalizing its demands for agrarian reform, Sincelejo clashed not only with the administration of Pastrana (1970-1974) but also with that of his successor, López Michelsen (1974-1978).

The political direction taken by ANUC-Sincelejo confirmed the government in its opposition to the organization. Its partisan political positions, indeed, had been forbidden by the law that established ANUC. Furthermore, ANUC-Sincelejo's positions were openly Marxist and opposed to the traditional parties. The Marxist orientation was in part a result of the evolution of peasant concerns, but, probably more significantly, it also resulted from efforts by Colombia's small leftist factions, none of which had a strong political base except the Communist Party, to establish one in ANUC. The meetings thus became an arena for contention among various leftist currents, and that in turn led to a number of internal divisions. One was over whether ANUC-Sincelejo should form a peasant political party or whether it should collaborate with other workers' groups in a common political enterprise. The organization has also divided between a group that favors "land for those who work it" and a more radical wing that looks to the ending of all private property ownership.

The decline of ANUC-Sincelejo is also a result of government repression. In addition to its open support for the Armenia line, the government canceled the legal status of the peasant associations linked to Sincelejo as a means of undermining the radical faction. In the numerous invasions and peasant protest demonstrations, some peasants and union leaders were arrested or killed by the police and armed forces. In 1974, the Sincelejo leaders sent a letter to incoming President López Michelsen asking him to explain his position on the peasant movement, but the President did not bother to reply. By 1979, participation in ANUC-Sincelejo had greatly declined, and membership in the organization had become semiclandestine.

Peasant organizations thus have a complex relationship to food-related policy. (1) Organized peasants are unquestionably important actors, but their influence is exerted through land invasions, demonstrations, and other activities outside the conventional channels of decision-making, although these actions have shaped the circumstances that led to "realistic" policy shifts by the elites. (2) The leaders of the Colombian system recognize the need for attention to peasant interests and participation by peasant organizations in official decision-making bodies. The government tolerates FANAL and ANUC-Armenia, while both federations are willing to shelve their emphasis on agrarian reform and accept the agenda of the government and large commercial producers. However, these concessions are based on the assumption that peasant representatives will agree with the basic presuppositions and policies of the larger landowners. In truth, the interests of peasants and landlords coincide only on increasing production; they differ emphatically on land distribution. (3) Organized peasants that oppose the assumptions of the government and hold radical political views are ejected from the dialogue and suffer various forms of repression.

Decisions on agricultural policy take place in private elite networks and by procedures in which peasants have traditionally not participated. Token peasant representation exists on the boards of directors of a number of institutions like INCORA (agrarian reform), IDEMA (marketing), and ICA (research and extension), but their views are apparently overshadowed by those of other groups like the producers' *gremios*. Furthermore, the peasant representatives come from ANUC-Armenia, which has been completely co-opted by the government. Finally, the institutions involved are principally policy-implementing rather than policy-making bodies.

The manner in which decisions about food-related matters are made in Colombia today can be illustrated by considering two quite different types of policy. The first involves credit, interest rates, exports, imports, and prices. Policy-making of this type is based on interaction between government institutions and the *gremios* of agricultural and livestock producers. The second is a broad general type of program like Integrated Rural Development (DRI) and the National Nutrition Plan (PAN), which were emphasized during the López Michelsen administration. The performance record, even the very existence of DRI and PAN, are associated with such factors as presidential leadership, technocratic entities in government, and international lending agencies. In neither policy area do peasants have any input into decision-making.

The Producers' Associations and Policy

The weakness of peasant federations contrasts sharply with the influence of the powerful associations of landlords, agricultural exporters, and others. Among the Colombian producers' *gremios*, the National Federation of Coffee Producers (FEDECAFE) has a special and exceptional role.[20] In their study of influences on agricultural policy-making, Bagley and Laun found that knowledgeable individuals mentioned FEDECAFE most frequently among private entities.[21]

One of the most important ways in which FEDECAFE differs from the other *gremios* is its intimate structural relationship with government. FEDECAFE was founded in 1927 by mutual initiative of the coffee producers and the government to defend and promote the industry domestically and abroad. Though formally a private entity, FEDECAFE is linked to the government in several ways: (1) The government collects taxes on coffee exports but then turns them over to the National Coffee Fund, which is administered by FEDECAFE. (2) Government officials represent 5 of the 11 members of the National Committee of FEDECAFE. The Colombian president and other government officials can veto or are authorized to approve certain activities of FEDECAFE. (3) The government delegates several functions to FEDECAFE, which include representing Colombia in the meetings of the International Coffee organization, regulating the activities of coffee processors, administering export licenses, and financing most of the development activities in the principal coffee-producing states.

Besides these more formal relationships, several presidents of Colombia in recent decades came from the coffee industry and its *gremio*. Finally, the financial resources of FEDECAFE are very extensive and include not only the Coffee Fund, which helps regulate prices and invests in the development of the coffee zones, but also an insurance company, banks, and most of the shares in the national shipping line.

Because of the vital importance of coffee to both the producers and government, their relationship is best described as one of coordination. The principal agent for FEDECAFE is the general director who maintains close personal relationships with the highest public officials, including the president. Since FEDECAFE has had only 2 general directors in the past 30 years, the person holding this office accumulates an enormous amount of power as representative of the most important sector of the national economy, his association with the dynamic state of Antioquia, the large sums of money he administers, and his network of personal contacts.

Despite the close relationship between the government and FEDECAFE, there is also a degree of what a reliable source describes as "permanent conflict" between them. The representatives of the coffee growers are concerned with defending the interests of their class while various government bodies focus more broadly on national matters, usually economic policies. Decisions generally result from complex discussions in which government agencies and FEDECAFE seem to hold comparable power. In cases of unresolved conflict, the nation's president makes the final decision.

The most frequently contentious issues between the other *gremios* and the government involve credit, prices, and imports that directly affect production. The nature of these relationships has not, to my knowledge, been studied, and information about them is obscured by broader questions like agrarian reform or integrated rural development.

A survey by Bagley and Laun asked three questions about agricultural policy, which were directed specifically to the National Front period and did not differentiate between types of policy. Instead, the questions focused on types of institutions: (1) In your opinion, what individuals and entities have influenced most actively economic policies in the agricultural sector? (2) In your opinion, which of the following government and international institutions has the major influence on national economic policies with reference to the agricultural sector? (3) In your opinion, which three institutions and associations of the private sector have the most influence on the economic policies of the government toward the agricultural sector?

The responses to these questions varied widely, probably in part because respondents had different policies in mind when they answered. The institutions they most frequently cited were the Ministry of Agriculture, FEDECAFE, the Agrarian Bank and other banks, and the Agency for International Development. Among the producers' associations, the overwhelming top three references were FEDECAFE, the Colombian Society of Agricultors (SAC), and the National Federation of Cattlemen (FEDEGAN).[22]

Besides FEDECAFE, which has its own mode of operation and is not organizationally related to the other *gremios*, Colombia has a number of associations representing producers of such commodities as cattle, rice, sugar, cotton, and milk. The strictly agricultural *gremios* are members of SAC, which represents the combined interests of the agricultural producers. On specific matters affecting its interests, a particular *gremio* tries to influence that policy alone, whereas SAC speaks for the generality of producers. Nevertheless, being a confederation, SAC is less powerful than the individual *gremios* and serves as an instrument for them.

SAC has an image of exerting considerable influence in Colombia because it agglutinates the interests of the large landholders, but several well-informed individuals noted in discussing its role that there are clear limits on its political power as an organization. One reason is that on certain key issues, like taxation in the 1950s and agrarian reform in the 1960s, the landowners had a great deal of influence, but they exercised it through alternative channels like the Congress or top officials in the executive branch. Second, SAC as an organization had not until recently taken the initiative in proposing agricultural programs to the government. Only when President Turbay Ayala came into office in 1978 did SAC, for the first time, make comprehensive suggestions to the new administration. Third, producers' influence on broad policies like agrarian reform has been reactive, modifying the proposals by criticizing them, or eventually blocking agrarian reform because they opposed changing the system of land tenure.

Similarly, a more recent policy like Integrated Rural Development did not originate from the producers nor did they help define it. Their influence has lain in their willingness to accept it, because they probably could have blocked it had they been opposed. In short, the *gremios* do not initiate policies, yet their veto power on rural policy sets the limits for acceptability. No major agricultural policy in Colombia can be implemented without their approval.

The relationship between the *gremios* and narrower, more specific policies like credit and imports can be ascertained by looking at the role of the Ministry of Agriculture, which most respondents to the Bagley-Laun questionnaire identified as most influencing policy. Although the Agriculture Ministry has considerable influence on policy, individuals familiar with its relationship to other government entities tended to emphasize to me that in the final decisions it was less significant than other institutions in its responsibility for overall national economic objectives. The Agriculture Ministry's influence comes from the fact that in important policy-making commissions it serves as the spokesman for producers' interests and draws up proposals on food policy. The Minister of Agriculture, however, does not make the final decision, and his suggestions are nearly always modified. In discussions with other ministers on policy-making bodies, the Minister of Agriculture usually lacks the sophistication and technical expertise to compete with the technocrats in other ministries and therefore cannot determine the outcome of decisions.

The institution that determines policy on credit, for example, is the Monetary Board (Junta Monetaria), presided over by the Minister of Finance and including other top officials like the

Minister of Agriculture and the Minister of Development (who represents industry), the Director of the Banco de la República, and the head of the Department of Planning. Each semester, the Minister of Agriculture draws up proposals for the volume of credit that is needed and for such allocations as seeds, machinery, infrastructure, and the various sectors. The advisers to the Junta Monetaria and the Banco de la República, who are usually outstanding young economists, also play an important role in preparing the proposals.

During the preparatory stage, the *gremios* are very active in pressing their case for a greater share of the available credit or increases in the allocation per hectare. The Minister of Agriculture receives numerous spokesmen, listens to them, and tries to incorporate their views and reconcile them in the proposals he draws up.

In the Junta Monetaria, the proposals are confronted with a different set of criteria, usually defended by the Minister of Finance. In this context strong personalities are significant in determining the outcome and the Minister of Development or the Director of the Banco de la República is often involved. According to the general pattern, though, the Minister of Finance is strong and certainly more so than the Minister of Agriculture, who is usually a political appointee. The Finance Minister examines the proposals from the point of view of his concerns with the availability of resources, the need to reduce inflation, or priorities in national production, and his tendency is to cut rural appropriations. A high official of SAC insisted to me that the most important institution for determining agricultural policy in Colombia is the Ministry of Finance because of its central role in the Junta Monetaria. He added that in this arena the SAC never gets all that it wants. Everyone with close knowledge of agricultural policy also recognized that the Junta Monetaria was a powerful but relatively little known force in agricultural policy.

Another very important decision-making forum is the Institute of External Commerce (INCOMEX). This entity, linked to the Ministry of Development, has the same ministerial composition as the Junta Monetaria, plus a representative of FEDECAFE and the Director of External Commerce. It, too, has advisers who are well-informed economists that play an important role in decision outcomes. INCOMEX makes policy in such key areas as the importation of fertilizers, machinery, or food items that are in short supply and must be purchased abroad. It is also in charge of exports and special relationships with the Andean Pact and other regional groups. On these questions, the Ministry of Agriculture similarly serves as a forum for other interest groups, which for fertilizers or food items may include processors as well as producers. Fertilizer quotas are worked out in a committee on fertilizers that includes agricultural producers, manufacturers of fertilizer, and technicians from the Ministry of Agriculture. The Ministry of Agriculture makes the summary proposal, but INCOMEX, influenced by factors like availability of foreign exchange and inflation, which are defined by the Ministry of Finance or INCOMEX itself, makes the final decision.

A third type of decision involves controlled food prices, particularly coffee, sugar, and milk.[23] The process in this case is different because of the large number of ministries whose interests are at stake. Here, too, the Ministry of Agriculture draws up the proposals after responding to the pressures from sugar and milk producers—who of course want the price to be as high as possible. The Minister of Agriculture presents his proposal to the Council of Ministers (the Cabinet), which makes the final decision. In this context, he must compete with the views of the Minister of Finance or perhaps the Minister of Health, who wants the price of sugar and milk as low as possible. The Ministry of Development has an important role in implementing the regulations and punishing violators through an agency under its jurisdiction called the Superintendency of Price Controls.

It should be clear from these examples that inputs into decision-making differ at the various stages of policy-planning. The influence of agricultural interest groups is exercised principally in the initial stages of the process, in the Ministry of Agriculture, although they may try to exert influence on any level. The Minister of Agriculture makes a preliminary reconciliation of the *gremio* pressures with productive necessities and hopes. The proposals that he formulates, however, theoretically represent the special concerns of the entire agricultural sector, which in turn have to be reconciled at the top with the interests of other ministers, especially the Minister of Finance.

The *gremios* also exert influence on subministerial levels. Whereas they do not get all that they want when the Minister of Finance is making decisions, they have a major impact on policy-implementing institutions like INCORA, the Agrarian Bank, ICA, and IDEMA through their representation on the board of directors and the acknowledged need for their cooperation if decisions are to be implemented. IDEMA establishes support prices for a few crops, but these are not generally considered to have been successful because they were set too low and were subsequently exceeded by market prices. On the local level, the larger producers usually enjoy special privileges from officials of the credit institutions or ICA because they are the important people in the area and naturally receive more consideration.

Rural Development and Nutrition Policies

The appearance of new programs of Integrated Rural Development (DRI) and Nutrition (PAN) in the 1970s helps round out the role of certain actors on food policy that have been previously discussed. Although the origin of these policies has not been studied and the programs are still too recent to evaluate, they stem from a variety of motivations and the interplay of several institutions.

Rural development and nutrition programs are not new in Colombia. The communal action programs, especially in the 1960s, attempted to coordinate on the local level the construction of schools and health posts in rural communities with programs to provide agricultural credit and technical assistance. INCORA carried out similar projects within the framework of the agrarian reform. Likewise, international nutritional organizations such as CARE and CARITAS have collaborated with the Colombian Institute of Family Welfare in carrying out food distribution and direct feeding programs since 1954. These efforts, however, have tended to lack comprehensiveness and sufficient resources to meet the needs especially of the poorer segments of the population.

Increased emphasis on rural development and nutrition programs can be ascribed to a number of causes. One was the assumption by the government that it ought to provide resources to the poorer classes of the population who were largely by-passed by agrarian reform. DRI was created to finance improvements in infrastructure, credit, and technical assistance for the smallest categories of peasant landholders, affected only minimally by earlier efforts. PAN provided food supplements and nutrition-related health improvements to communities with more than ordinary poverty and malnutrition. (Malnutrition in Colombia is concentrated among women and children, especially families of landless agricultural workers. Studies between 1963 and 1968 suggest that about two-thirds of the children in the sample suffered from malnutrition.[24]) DRI was integrated into the nutrition program: it assumed that increased food production among smallholders would improve their own consumption and increase the availability of the most common food items in the market.

A second influence has been the development of an international consensus in the United Nations, Latin American regional organizations, and international lending institutions that the problems of the most neglected segments of the population must be confronted. As a consequence, a number of countries besides Colombia have also initiated fresh approaches to rural development and nutrition. Loans and technical assistance have been available for them from national and international donors.

A third important factor was a change in the presidency, bringing into office Alfonso López Michelsen in 1974. Although López did not favor agrarian reform, he had long been identified with the more progressive wing of the Liberal Party. The principal guidelines of his administration were shaped by the slogan "To Close the Breach," implying some redistribution of income and a reduction of the gap between rich and poor.

The combination of internal circumstances, international influences, and domestic political concerns coalesced to produce new food policies in the 1970s as they had in the 1950s and 1960s. The rural lower classes, the major intended beneficiaries of these programs, had no influence at all in promoting them or designing them except by expressing their dissatisfaction outside decision-making channels. Organized peasant groups, including ANUC-Sincelejo, accepted the policies once they were established though they criticized the unwillingness of the government to include with them an agrarian reform. The large landholders offered no resistance because they favored programs to assuage rural tension; the emphasis on integral development, moreover, corresponded to their own sympathy for capitalist modernization of agriculture. The producers' *gremios* have praised DRI, assuming that the larger-sized peasant farms (5 to 20 hectares) will join them as capitalist producers, while the smallest ones (less than 5 hectares) may abandon, sell, or neglect their holdings and join the available labor pool.[25] (In certain regions and seasons of the year Colombia has a shortage of agricultural labor because of urban migration and the exodus of between one and two million Colombians to work in Venezuela.)

Some Colombians complain that their government has been a willing instrument of international lending agencies which in fact determine national policies. Closer examination of DRI and PAN will help clarify the relative impact of external and internal influences on food policy decisions.

There is no question that Colombia has been one of the leading beneficiaries of international loans since the 1950s. The impact of the U.S. Agency for International Development (AID), the International Development Bank (IDB), and the World Bank has coincided with the expansion and increased power of technocrats in the executive branch of the Colombian government to plan and administer the programs. The presidency has increased its strength at the expense of Congress and the technocrats now have far more impact on policy than the traditional or "political" bureaucracy. The institution whose star has most clearly risen is the National Planning Department (DNP), which formulates the plans, in consultation with technicians of the international agencies, that are required to obtain loans.

Over the years Colombia has developed great skill at securing loans. It was, for example, the first country to submit an acceptable plan to the Alliance for Progress. The image of Colombia as a country first in line with its plan when money is available for certain types of development has nourished the view that international lending agencies play a decisive role in the country's policies. The most sophisticated form of this argument is found in the writings of the Colombian political scientist Fernando Cepeda Ulloa. The behavior of Colombian technocrats, he says, is determined by:

> *...the aspirations, political conceptions, and philosophy of development of the international agencies that have been inspiring the development of Colombia in the last 25 years.... These technocrats have been prisoners of the policies of financing designed by the international agencies.... Our plans of development were effective when they were accommodated to those financial possibilities and ineffective when they departed from them.*[26]

Cepeda Ulloa's chief concern over this tendency stems from the emergence of an elite dependent on international guidance and the consequent loss of political influence and control over policy-making:

> *In the measure that technical criteria become more important, the process of sectoral or global planning acquires more relevance, and there is less opportunity or necessity to obtain a firm popular support for specific policies. For this reason, political initiative at the popular level and the response to this initiative become less and less relevant in the process of formulating public policies.... The process of planning has occurred at the margin of the political parties, social forces, public opinion, and most seriously, the national representation in the Congress of Colombia.*[27]

Although in the 1960s and early 1970s Congress provided a forum for discussing agricultural policies such as agrarian reform, Congress had no influence at all on the DRI and PAN proposals, making them ideal examples of the interplay of international lending agencies and the technocracy, supported by the president. López Michelsen made use of emergency powers granted him by the Congress to decree his program.

Although international lending agencies are undeniably significant, it is more accurate to recognize a coincidence of interests between them and Colombian policy-makers rather than to ascribe the policy merely to outside pressure or influence. The lenders want to promote certain kinds of programs, but the Colombian technocrats, operating with their own perceptions and circumstances, also have an interest in them. The need for compensatory rural programs because of the neglect of agrarian reform, as well as elite perceptions about unrest and its solution are purely Colombian factors. Furthermore, Colombian technocrats are influenced by the same international discussion and consensus on development strategy that have reshaped the orientation of the lending agencies.

DRI's origins were in the early 1970s. The planning office of the Ministry of Agriculture was engaged in discussions with the decentralized agricultural agencies like INCORA, ICA, the Agrarian Bank, and IDEMA about expansion of rural development efforts to the lower classes. Some of the obvious needs were a more adequate supply of credit (which under the agrarian reform had gone mainly to large and medium producers), a more effective extension service (ICA was very limited), mechanisms for improving the marketing of products (IDEMA was notably inept), and resources for infrastructural improvement. Conversations were going on at the same time with the World Bank and the IDB about financing. Already under the Pastrana administration, an FAO mission was contracted with the help of international financing to make a study of the regions with large concentrations of *minifundios* (very small farm units) which would be the principal beneficiaries of DRI investments.

PAN also had its beginnings in discussions under the Pastrana administration. International circumstances played an important part when, as a result of the worldwide grain shortage of 1972 and 1973, various international donors to the existent feeding programs, including the United States government, announced that in the proximate future they would shift their activities to countries with more serious food and nutrition problems than Colombia. In 1972, the Colombian government appointed an interministerial National Committee on Food and Nutrition Policies, which began to function early in 1973. Over a period of many months of discussion, the basic outlines of the program emerged. Officials of AID participated in and contributed to the discussions but, according to participants in them, AID did not impose its views. Both AID and Colombian officials favored a multisectoral program; its precise form was the result of mutual and multiple contributions, both personal and institutional. In July 1973, the Plan was presented at a meeting of the Andean Pact countries; and in December 1973, a national nutrition conference was held in the city of Palmira, Valle, to get the reactions of medical, nutritional, educational, and other experts to the proposals.

The Pastrana administration had been moving in the direction of what were to become DRI and PAN, but leadership was weak. The crystallization of DRI and PAN required a new president, López Michelsen, and a new institutional setting, the National Planning Department. López was looking for issues to emphasize during his campaign and administration and, given his political orientation, DRI and PAN were natural for him. When he came into office, the two programs were moved out of the agricultural and nutrition frameworks and turned over to the DNP.

The DNP has had a decision-making role since the 1950s, advancing new policy initiatives, drawing up proposals, and securing outside financial support. Individuals familiar with the institution nevertheless emphasize that its power is contingent on that of the president and his interests. Colombia's presidents have always played an important role in policy, and even relatively weak presidents have been prominent national leaders with long experience in public life. They develop their political orientations through speeches and writing before becoming president, and once in office they try to give a new direction to national policy. This is evident in the Alliance for Progress reformism of Lleras Restrepo, the "Four Strategies" of Pastrana, and López Michelsen's "Closing the Breach." Certain individuals in DNP may have had an exceptional role— Miguel Urrutia, its head under López, or Eduardo Weisner under Turbay Ayala—but they too have only the power and influence the president gives them.

It is too early at this point to evaluate how successful PAN or DRI have been. Nor is it clear how the inauguration of the Turbay administration (1978-1982), representing the more conservative wing of the Liberal Party, has affected the two programs. What is clear is that rural poverty, rural landlessness (see Tables 2 and 3) and malnutrition continue to be serious problems which DRI and PAN must try to alleviate.

Conclusion

The preceding discussion indicates that food policy decision-making in Colombia takes place fundamentally in the executive branch of the government. This conclusion corresponds to the results of the Bagley-Laun questionnaire, which indicated that "the most influential institution in economic policy-making in general was the Executive branch of the national government with 31.5 percent of all weighted mentions."[28] Examination of production-oriented policies like credit and imports, plus recent programs like DRI and PAN, has tried to clarify *how* these decisions are made and by which entities in the executive branch. Comparing evidence supplied by Bagley and Laun's studies with more recent experience, this analysis also suggests that the role of the president and the technocrats is becoming more significant, relative to Congress and the *gremios*.

This is not to say, however, that Congress and the *gremios* could not once again affect policies more directly. Congress has recently debated legislation for establishing price stabilizing funds for other commodities besides coffee; and the presentation by SAC of a comprehensive agricultural program to President Turbay Ayala in 1978 is an admitted attempt by that organization to coordinate its resources to exert more influence. Still, the growing dominance of the presidency and the technocracy in Colombia corresponds to similar tendencies in several other Latin American countries that have experienced a conflict between the imperative to move ahead on development programs and the tendency of political and interest group differences to immobilize channels of action.

Producer interest groups have their greatest influence on Colombian food policy by defining the outside limits or parameters within which it takes place. Coming from backgrounds linked with large landholders and commercial producers, most recent Colombian presidents have not questioned these limits. The consensus of both favors more intensive use of land, increased production, modernization within a capitalist framework, and no agrarian reform. The president and his advisers on the one hand and the producers on the other are symbiotically linked: the former cannot advance agricultural development without the firm collaboration of the large private producers; the landowners depend on public resources and policies to increase production.

Although the president, the technocrats, and the producers agree on their overall orientations, conflicts do emerge on specific policies. Differing concerns exist among the producers themselves, and between them and other entities of government. Although the government wants to satisfy the producers, it must also respond to other realities like the need for industrialization, demands of urban consumers, multiple demands on limited resources, and the resolution of ongoing economic problems like inflation.

International lending agencies have significantly encouraged certain types of policy in Colombia, participated in determining their form, and provided financing. Nevertheless, these activities take place in contacts with Colombian planners whose basic orientation is also influenced by Colombian problems and policy interests.

Colombia's close linkage to international lending agencies in the past 20 years has not been an expression of conformism, but rather a form of self-interest. The policy emphases of AID, IDB, and the World Bank have been reasonable proposals for development that Colombians were able to utilize. As a result, Colombia has greatly improved its infrastructure as well as its social programs. With its ample foreign exchange earnings, it has contracted these loans without suffering a serious debt burden. Furthermore, the insistence of the lending agencies on forming new, more responsible, and technocratically oriented institutions to carry out many of the programs they financed has strengthened the quality of personnel and administrative skills. Sizable loans from the international agencies have helped Colombian presidents initiate and expand programs useful to national development, at a time when acquiring funds from the Congress might have been difficult or impossible.

The obvious negative feature of the Colombian pattern of agricultural decision-making is the marginalization of political parties and peasant organizations. The parties do not represent the interests of the country's diverse social groups, and the dominating elites who control the parties often prefer to express their views by other means. At any given moment, only a few hundred people in Colombia are important contributors to decision-making. Though they represent different political views, interests, or institutional commitments and may be in conflict, at the same time they know each other and are in frequent contact through meetings and phone conversations.

The poorer farmers—peasants, renters, sharecroppers, and *obreros*—do not have this compensation. Their chief interest, agrarian reform, is excluded from the agenda, and their organizations do not play a substantive role in the negotiations on the more specific policies that affect production. Similarly, despite impressive gains in Colombian food production (including export crops), malnutrition remains a serious problem and the cities have *not* provided adequate employment for many of the peasants who have been pushed off the land by the expansion of commercial agriculture. Without neighboring Venezuela as a source of employment for up to two million illegal Colombian workers, the problem might be even more serious.

The DRI and PAN programs illustrate the interrelationship of domestic and international actors. The poorer rural classes did not lobby for them because they have no standing or channels to do so. Rather, the two programs were worked out in consultations between the president's technical advisers and officials of international lending agencies. The large-scale producers, though they do not benefit from DRI and PAN, approve the programs because they do not challenge their position and, in certain respects, meet their concern for rural stability and the strengthening of a capitalist agricultural system.

Colombians generally admit that DRI has helped reduce tension in the countryside, with an assist from the government's policy of repressing rural dissent. Many smallholders will surely benefit from DRI and increase their production, although most analysts believe the smallest and weakest among them will probably continue to abandon farming. Young people especially are discouraged about their inability to earn a satisfactory living from small plots and seem likely to continue entering the urban and rural labor forces at the bottom levels. Thus rural employment, nonskilled or semiskilled urban employment, and improved nutritional standards among the poorest 40 percent, rural and urban, are likely to remain national problems for the future.

(August 1980)

NOTES

1. The classic study of agricultural policy by Albert O. Hirschman, *Journeys Toward Progress: Studies of Economic Policy-Making in Latin America* (New York: Twentieth Century Fund, 1963), is very helpful for the period through approximately 1962. The 1973 laws and their background in the agrarian reform are analyzed in Bruce M. Bagley and John Laun, "Political Power and Agricultural Policy in Colombia: A Case Study of Laws 4 and 5 of 1973," mimeograph (Bogotá: Department of Political Science, University of Los Andes, 1977). The study by Bagley and Laun uses as part of its methodology a questionnaire on institutions and individuals influential in decision-making.

2. Agrarian reform is the one aspect of Colombian food policy that has already been extensively studied. Cf. Hirschman, *Journeys Toward Progress*, pp. 141-158; and Ernest A. Duff, *Agrarian Reform in Colombia* (New York: Praeger, 1968). On its decline, cf. Bagley and Laun, "Political Power and Agricultural Policy in Colombia."

3. Thomas G. Sanders, "PAN: A Description of the Colombian National Nutrition Program," Office of Nutrition, U.S. Agency for International Development (July 1980).

4. Some studies are projected. Jonathan Hartlyn is writing a dissertation in the Department of Political Science, Yale University, on the role of interest groups in policy-making. The Fundación para la Educación Superior y el Desarrollo (FEDESARROLLO), a research institute in Bogotá, is also in process of developing a multifaceted research project on how decisions are made.

5. On the Colombian political system, cf. Robert H. Dix, *Colombia: The Political Dimensions of Change* (New Haven, Conn.: Yale University Press, 1967); John D. Martz, *Colombia: A Contemporary Political Survey* (Chapel Hill, N.C.: University of North Carolina Press, 1962); and James Payne, *Patterns of Conflict in Colombia* (New Haven, Conn.: Yale University Press, 1968). Albert Berry, ed., *Coalition Politics in Colombia* (New Brunswick, N.J.: Transaction Books, forthcoming). An excellent recent essay on Colombian political structure, emphasizing its "consociational" dimension is Alexander Wilde, "Conversations among Gentlemen: Oligarchical Democracy in Colombia," in Juan J. Linz and Alfred Stepan, eds., *The Breakdown of Democratic Regimes: Latin America* (Baltimore, Md.: Johns Hopkins University Press, 1978), pp. 28-81.

6. Hirschman, *Journeys Toward Progress*, pp. 109-113; and Salomon Kalmanovitz, *Desarrollo de la Agricultura en Colombia* (Bogotá: Carretas, 1978), pp. 25-27. The latter book deserves citation as the best study of the economic development of Colombian agriculture.

7. Ibid., pp. 64-65.

8. Ibid., pp. 38, 185-86; and Table 2 of this *Report*.

9. Hirschman, *Journeys Toward Progress*, pp. 116-35.

10. Kalmanovitz, *Desarrollo de la Agricultura*, pp. 152-162; and Table 1 of this *Report*.

11. Cited in Duff, *Agrarian Reform in Colombia*, p. 54.

12. On Lleras Restrepo's views, cf. Carlos Lleras Restrepo, *El Cambio Social* (Bogotá: Agra, 1967).

13. Juan Sebastián Betancur Escobar, "La Política Agraria y el Plan Desarrollo," in *Coyuntura Económica* V (July 1975), p. 94.

14. *El Cuatrenio de la Transformación Rural* (Bogotá: Ministerio de Agriculture, 1970), p. 37.

15. Bagley and Laun, "Political Power and Agricultural Policy in Colombia," p. 52.

16. Laughlin Currie, *Ensayos sobre Planeación: Introducción a una Teoría de Desarrollo Conocido como Operación Colombia* (Bogotá: Tercer Mundo, 1963), p. 72.

17. The Chicoral Pact is found in Oscar Delgado, ed., *Ideologías Políticas y Agrarias en Colombia* (Bogotá: Tercer Mundo, 1973), pp. 141-45.

18. The juntas of communal action are discussed in Bruce M. Bagley and Matthew Edel, "Popular Mobilization Programs of the National Front: Cooptation and Radicalization," in Albert Berry, ed., *Coalition Politics in Colombia*.

19. The development of ANUC and the evolution of the Sincelejo line are discussed in ibid., and in Bruce M. Bagley and Fernando Botero Zea, "Organizaciones Contemporáneos en Colombia: Un Estudio de la Asociación Nacional de Usuarios Campesinos (ANUC)," in *Estudios Rurales*, No. 1. Cf. also the interview with Ricardo May Cabrera, Secretary General of ANUC, in *Alternativa*, No. 6 (May 1-15, 1974), p. 4; and *Alternativa* in general.

20. On FEDECAFE, cf. FEDESARROLLO, *Economía Cafetera Colombiana* (Bogotá: Fundo Cultural Cafetero, 1978), esp. pp. 377-463.

21. Bagley and Laun, "Political Power and Agricultural Policy in Colombia," p. 148.

22. Ibid., Appendix I, pp. 144, 148.

23. On the role of support prices, cf. Kalmanovitz, *Desarrollo de la Agricultura en Colombia*, pp. 136-83; and Dragoslav Avramovic, ed., *Economic Growth of Colombia: Problems and Prospects* (Baltimore, Md.: Johns Hopkins University Press, 1972), pp. 252-54.

24. Tomás Uribe Mosquera et al., *Nutrición* (Bogotá: Departamento Nacional de Planeación, 1979), p. 28.

25. The data from Kalmanovitz (p. 54) indicate a decline in the number of small farms between 1960 and 1971. See Table 3 of this *Report*.

26. Fernando Cepeda Ulloa, "Planeación y Dependencia," in *Ensayos Políticos* (Bogotá: Universidad de los Andes, 1976), pp. 7-8.

27. Ibid., p. 5; and "La Influencia dé las Agencias Internacionales en el Proceso de Desarrollo de Colombia, 1950-1974," ibid., p. 172.

28. Bagley and Laun, "Political Power and Agricultural Policy in Colombia," p. 128.

Peasants, Landlords and Bureaucrats
The Politics of Agrarian Reform in Peru

Howard Handelman

The bus from the state department capital of Cuzco into the Anta Valley carried us only as far as the town of Izcuchaca. There I disembarked with a small group of smartly dressed men and women in sport jackets and pant suits, carrying their cameras and new suitcases. All of us were traveling to the fifth national congress of the Confederación de Campesinos del Perú (CCP—Peruvian Peasant Confederation) which was being held in the small Andean village of Chacán. I soon learned that my Peruvian, apparently middle-class companions were not reporters nor an urban support group, as I had imagined, but a delegation to the peasant congress representing small landowners from the coastal state of Lambayeque. Their dress and manner sharply distinguished them from the local highland villagers (wearing Indian ponchos and sandals) who chatted in Quechua and watched the "gringos" (a term applied equally to foreigners and to whites from the Peruvian coast) with bemusement. It was the first time that any of the Lambayeque smallholders had ever been to the sierra (Andean highlands) and they excitedly snapped their cameras as a pack of llamas and alpacas appeared down the street. Surprisingly, they seemed less comfortable than I did among their fellow countrymen, their fellow *campesinos* (peasants) whom they were about to join in a national congress.

Starting out on the road toward Chacán, we waved down a truck filled with Indian villagers from the southern sierra province of Andahuaylas, a hotbed of peasant unrest in recent years. The Andahuaylans (some 50 of them) were delegates from various peasant communities who had traveled for over 15 hours along rough dirt roads in order to attend the congress. As we rode they sang a Quechua protest song to the haunting accompaniment of Andean flutes. On the outskirts of Chacán we were stopped by a group of young "security guards" armed with pitchforks and clubs. Technically the CCP is not a legal rural organization, not having received government recognition as a peasant interest group under the terms of the 1974 rural mobilization law. Consequently, although the government was not interfering with the congress, these village guards were demanding credentials from delegates and letters of invitation from reporters and support groups in order to keep out "spies."

Once a major spokesman for peasant discontent during the early 1960s (with the support of urban leftists), the Confederación de Campesinos became inactive in the middle of that decade owing to government repression of radical peasant organizations as well as limited concessions to campesino demands (through a state community development program and scattered agrarian reform). In 1974, the independently Marxist Vanguardia Revolucionaria helped revive the CCP at the Confederation's 4th National Congress. The organization's current support among landless peasants and smallholders in the northern state of Piura, the southern highland regions of Cuzco, Puno, and Apurímac, and (to a lesser extent) the coastal states of La Libertad and Lambayeque reflects the widespread discontent of many Peruvian peasants who haven't received the anticipated benefits of the military government's sweeping 1969 Agrarian Reform Decree.

Today, the CCP is one of Peru's two major national peasant organizations. The other one, the Confederación Nacional Agraria (CNA) was established in 1973-74 under the auspices of the "revolutionary" military government for the purposes of combating landlord resistance to agrarian reform. In mid-1978, the government dissolved the CNA at the national level after its top-ranking officers had proven to be too radical and independent for the tastes of a more conservative military regime. Avelino Mar, Secretary General of the CNA at the time of its official dissolution, and other CNA leaders attended the CCP Congress at Chacán in a move toward greater organizational cooperation.

The unfolding of the CCP's National Congress (the first since the organization's 1974 revival) suggested both the peasantry's potential for independent mobilization and the many obstacles still impeding the growth of a national campesino movement. Throughout the week one might observe various unresolved divisions within the confederation both at the rank and file and at the leadership level.

From the time I and the Lambayeque delegates first climbed onto the truck carrying the Andahuaylas villagers, I was struck by the cultural, economic, and linguistic differences between the *mestizo*, Spanish-speaking smallholders from the coast and the Quechua (or Aymara) highland campesinos. Later, at the congress itself, in talking with delegates from Peru's northern coast, I learned that many of them were small and medium farmers raising cotton, corn, and sorghum. Their principal concerns were for higher commodity prices and more government loans—rather modest, indeed petit-bourgeois, demands. Villagers from Cuzco and Puno—mostly cultivators of small subsistence plots or landless agricultural workers—were far more bitter and spoke angrily of their failure to receive land under the agrarian reform.

Within the highland peasantry, further divisions existed. I approached a group of villagers from the barren *altiplano* (high-altitude grazing region) of Puno, near the Bolivian border. They carried with them an Aymara flag which they proudly said they had successfully defended (at the cost of several peasant lives) two and one-half years earlier against attempts by local police to lower their banner and raise the Peruvian national flag. Their community leader expressed anger over the fact that bilingual Quechua-Spanish courses were being offered in many Cuzco villages while Aymara instruction was allegedly unavailable in Puno. Another Aymara-speaking peasant spoke contemptuously of the Quechuas' alleged lack of work ethic: "If we had land like this," he said, pointing to the rich Anta Valley, "we wouldn't be going hungry."

After the hundreds of delegates had registered at Chacán, they assembled (by region) for a march back to Izcuchaca where an opening rally was to be held. I learned that Izcuchaca had recently been the site of a land seizure in which peasants from a nearby *comunidad* (a peasant village of smallholders with communal grazing land and work forms) had invaded a neighboring agrarian reform cooperative (i.e., an expropriated hacienda) because they had not been afforded the use of co-op land. Thus the rally site itself epitomized the differences between the beneficiaries of agrarian reform and the many villagers whose exclusion constitutes yet another impediment to peasant unity.

As we marched into Izcuchaca's main square, a battle for position was developing between supporters of Vanguardia Revolucionaria (the CCP's largest ideological faction) and those of the Partido Comunista Revolucionaria (a pro-Chinese faction), who also attempted to outshout each other with their respective slogans. The following day, I watched peasants at the congress gather in little circles in the fields outside Chacán listening to VR and PCR spokesmen engage in esoteric debates about the present state of the military's "bourgeois revolution" and the proper attitude of the radical left toward the announced transition to elected civilian government.[1] To my amazement, some of the village leaders to whom I spoke expressed interest in these highly ideological debates so far removed from the immediate needs and problems of the peasantry. Yet, most delegates seemed to view them as a necessary, if confusing prelude to discussion of the real issues. Moreover, in my conversations with delegates and with onlookers (from nearby Cuzco communities) at the Izcuchaca rally, I observed that the CCP leader with the highest degree of popularity among the campesinos was Hugh Blanco, a Quechua-speaking, Cuzco Trotskyist (elected to the Peruvian Constituent Assembly in the June 1978 national elections) whose Partida Socialista (PST) party had little influence in the Confederación de Campesinos.[2]

Cleavages such as these mute, but do not silence the voice of rural discontent. Since the outbreak of widespread peasant land invasions in the early 1960s and the accompanying growth of campesino federations, the Peruvian government has no longer been able to ignore the voice of its rural tenants, smallholders, and agricultural workers. On June 24, 1969, General Juan Valasco Alvarado, President of the revolutionary military regime that had seized power the year before, announced the promulgation of the most sweeping land reform program in Latin America since the Cuban Revolution. "Peasants," he declared, "the landlords will no longer eat from your poverty."[3]

Unlike most of the Latin American agrarian reform laws issued in the 1960s, Peru's land redistribution was carried forth vigorously. By 1978, the traditional Andean haciendas and coastal plantations that had long dominated Peruvian agriculture were no longer in the hands of an all-powerful rural oligarchy. The political and economic power structure of the countryside and, hence, of the nation as a whole had been significantly altered. The CCP's resurgence in recent years and the CNA's growing militancy, on the other hand, indicate that, whatever its accomplishments, the government has failed to bring tranquility to the countryside. The search for explanations for this failure must begin with an appreciation of conditions before the rural reform, then proceed through the complex of political forces unleashed by changes in land tenancy, the mobilization of rural interest groups, the government's reaction, and the effects of all these changes on agricultural production.

Peru's Traditional Rural Structure

Understanding Peruvian rural change since 1969 requires a brief review of land tenancy and rural power relationships prior to the reform. Underlying both is the distinction between coastal and highland agriculture. While both regions were dominated prior to 1969 by large latifundia, they differed in terms of their production methods, labor relations, and markets.[4]

The Coastal Plantations. Most of the thin strip that runs along the Pacific coast of Peru is a barren, uninhabited desert. However, in a series of valleys, irrigated by rivers that flow down from the Andes, are located the nation's largest cities (including the capital, Lima, and its twin city, Callao), half the country's population, and its capital-intensive, modern agricultural sector. The coast, particularly Lima, has long been the center of political and economic power as well as the seat of its *criollo* (white, Spanish) culture. In addition, the vast sugar plantations of La Libertad and Lambayeque have produced Peru's principal twentieth-century agricultural export.[5] The sugar barons diversified their capital into banking, commerce, and industry as well. Indeed, "coastal sugar and cotton exporters had formed the dominant sector of the country's ruling coalition during the twentieth century" prior to the agrarian reform.[6] Smaller coastal haciendas (generally 50-250 hectares) also produced the rice, milk, and much of the vegetables for urban markets.

Coastal export agriculture (particularly sugar) has been characterized by a relatively high degree of capital investment (most notably in irrigation) and mechanization. Labor was provided by a permanent, salaried work force (supplemented by seasonal workers hired during the harvest) which generally manifested a working-class, rather than peasant orientation. Employer-worker relations were often impersonal and sugar workers, along with many other coastal agricultural workers, were organized into unions for more than two decades prior to the agrarian reform. Most of the unions—including the powerful Federation of Sugar Workers—were dominated by APRA, a reformist, populist political party which has brought together middle- and lower-class voters into Peru's largest political organization.[7] Worker demands were usually moderate and trade-unionist: higher wages and fringe benefits, better working conditions. The union members were little interested in agrarian reform and, indeed, when Peru passed its first, modest, land redistribution legislation in 1964, APRA insisted that the coastal plantations be excluded. During the late 1960s and early '70s, APRA union leadership was successfully challenged by leftist insurgents in some of the Lambayeque plantations.

The Traditional Highland Hacienda. If the Peruvian coast is the center of criollo culture and power, then the sierra is the repository of Indian and mestizo tradition. Two-thirds of the *serrano* population is rural and for the majority of highland peasants, Quechua is the primary language. Less than half the region's adults are literate.[8] Following agriculture, the sierra's major economic sector is mining, the source of Peru's most important exports. Primarily, however, the highlands have served the coastal cities as an "internal colony," a source of cheap labor, food, and raw materials.

Unlike the coastal plantations, the Andean haciendas have usually produced food for nearby local markets or for Lima-Callao. Except for some wool trade, Andean agriculture has not served the export market. Particularly in the south, huge latifundia—devoted primarily to raising sheep, llamas, alpacas, and (at lower altitudes) cattle—have dominated the landscape. In Ayacucho, Huancavelica, Puno, and Apurimac, haciendas of over 500 hectares held 59-82 percent of the agricultural-livestock land area. In Puno there were 11 latifundia of 20,000 to 46,000 hectares.[9] These vast estates coexisted with thousands of peasant communities (Indian or mestizo) often containing both agricultural plots (each usually under 5 hectares in size) and communal grazing land. Peasant agriculture, along with many medium-sized haciendas, was primarily devoted to the production of corn, potatoes, grains, and vegetables either for subsistence consumption or for sale on the domestic market.

While the sierra's *hacendados* (hacienda owners) lacked the national economic and political power of the coastal oligarchy, they had great influence at the local level. Within the hacienda itself, they exercised considerable power over the neofeudal peons (called *feudatarios*) well into the middle of the twentieth century. In return for usufruct of a small plot of land and, perhaps, limited grazing rights, the hacienda feudatario rendered virtually unpaid labor on the hacendado's land for as many as 150-200 days per year. Often the peons were required to sell their crop surplus to the landowner at below-market prices. Resistance to this neofeudal system was reduced by the peasants' dependence on their "*patrón*" for tools, loans, aid in times of medical need, protection from the police, and support of fiestas.[10] Traditionally, the feudatarios were kept isolated from "potentially dangerous" contacts with outsiders, with peasants on other haciendas, and even—through the hacendados' tactics of playing peons off against each other—from other peons on the same estate. Peasants who became "trouble-makers" could be expelled from the hacienda or worse. Thus, until the 1960s, unionization of hacienda workers was extremely rare and even during the outbreak of peasant unrest in the early part of that decade, the feudatarios were less mobilized than village smallholders.[11]

The majority of highland peasants were not bound to the haciendas: most were village *minifundistas* (owners of small plots), and a few were landless, transient laborers. The peasant smallholders—particularly those belonging to organized communities—were more independent than the hacienda peons,[12] although they too were subject to the economic and political power of the hacendado and his allies among the merchants, police, and local officials. This was particularly true of those minifundistas who were forced to supplement their income by working part of the year on nearby haciendas.[13] Finally, the independent peasant communities often were faced with long term encroachments by the haciendas on the villages' communal grazing land.

Throughout Peruvian postcolonial history, there have been periodic peasant rebellions; between 1921-22 (a period of particular unrest) and the mid-1960s, there were 33 recorded revolts in the southern sierra alone.[14] With some notable exceptions, however, these hundreds of scattered *jacqueries* and other "primitive rebellions" were generally isolated movements, lacking organization, ideology, or well-defined goals and, hence, were easily crushed.

Land Tenure

Although the types of agricultural systems predominant in the two regions differed, both the more capitalistic coastal farms and the more feudal Andean highlands were marked by the heavy concentration of land in a small number of units. The results of a 1972 national agricultural census revealed the extent of such concentration. (Although the census was conducted three years after the promulgation of the agrarian reform, the data reflects the pre-reform situation, both because hacienda expropriation had scarcely begun in the highlands, and because those estates already expropriated on both the coast and the sierra were retained as cooperative production units and were not broken up.[15]).

Smallholders (0-5 hectares) accounted for 80 percent of the agricultural units on both the coast and the sierra, but held less than 10 percent of the land (see Table 1). Indeed, over half of those *minifundia* were under one hectare and accounted for a total of only 1.1 percent of coastal and 0.8 percent of Andean farmland. At the other end of the scale, 62 vast coastal estates owned nearly half of that region's agricultural area. Although the figures for the Andes appear even more extreme, the data are slightly misleading: the census table did not distinguish between haciendas and communally owned (village) pasture lands, thus failing to reveal that, of the 15 million hectares of sierra land concentrated in units of over 200 hectares, some 6 million actually belonged to 1,235 village pastures.[16] Consequently, large highland haciendas (the 5,400 estates that held over 200 hectares) actually controlled "only" 46 percent of

Table 1

Land Concentration on the Peruvian Coast and Highlands

Size of Unit	Coast: Number of Units (%)	Coast: Area (Hectares) (%)	Highlands: Number of Units (%)	Highlands: Area (Hectares) (%)
0-5 hectares	135,651 (79.9%)	176,568 (9.5%)	884,209 (81.7%)	1,168,998 (6.5%)
5-50	31,416 (18.5%)	350,657 (19.0%)	179,937 (16.6%)	2,049,688 (10.6%)
50-200	2,271 (1.3%)	218,589 (11.8%)	11,386 (1.1%)	1,086,931 (5.5%)
200-2,500	486 (0.3%)	245,944 (13.4%)	5,741 (0.5%)	3,455,042 (17.9%)
over 2,500	62 (0.05%)	855,513 (46.3%)	894 (0.1%)	11,530,803 (59.5%)
Total	169,886	1,847,271	1,082,167	19,291,462

Source: *II Censo Nacional Agropecria* (Lima: Oficina Nacional de Estadística y Censos, 1975).

the region's land. Even with this adjustment, latifundia domination was almost as great in the highlands as on the coast.

Socioeconomic Changes

The first organized challenges to this strongly hierarchical rural structure originated, not surprisingly, on the more capitalistic, modernized coast. The plantation workers' greater hispanization, higher levels of education and literacy, and greater exposure to the national political culture (as compared to the Andean peasantry), coupled with the impersonalized labor relations of corporate agriculture, produced a certain degree of working-class consciousness. During the 1940s, with the help of APRA party sponsors, coastal plantation workers were able to secure government recognition of their unions, the right to collective bargaining, the minimum wage, and the few social security benefits afforded by Peruvian law. While certain union locals became militant at times during the next two decades, and while strikes were sometimes violent, the region's agricultural workers—tempered by the APRA unions' moderate orientation—never challenged the national political system or the bases of the country's agricultural structure.

In the sierra, the peasantry's organizational capability was impeded by its frequent lack of familiarity with the Spanish language, low levels of literacy and education, and relative isolation from national politics. Through the 1950s and 1960s, however, some socioeconomic changes were taking place that eventually led to widespread mobilization within the Andean peasantry. When the campesinos finally confronted the culture of domination and oppression, their challenge was far more dramatic and militant than that of the coastal workers.

The spread of rural education in the highlands gradually produced a pool of politicized (often radical) students and teachers who provided the leadership skills and sophistication for incipient peasant organization. Increased literacy and the spread of mass communication (mainly radios) also brought villages in contact with the Peruvian political system. Finally, the extensive migration of Andean villagers to urban centers (Lima and highland Department capitals) brought the migrants into contact with unions, political parties, and other agents of change. Those who returned to their communities to visit or to live provided new ideas and skilled leadership.

These forces of modernization were, of course, spreading throughout Latin America and the Third World during the postwar decades. In Peru, however, they were augmented by factors particular to that country. In the central highland states of Pasco and Junín, the expansion of mining and metal refining drew large numbers of peasants from neighboring communities into the modern work force. Through the miners and metal workers' union, these former peasants established a network of contacts with other villagers in the region. They also learned basic political and organizational skills and developed greater political consciousness and militancy.

All these factors were superimposed on a steady decline in the economy and philosophy of the traditional hacienda system. Increasing numbers of hacendados, anxious to leave the countryside for the greater attractions of the city (or pressured to do so by their children), managed their property as absentee landlords, thereby reducing control over their peons. Others sold their estates—in part or entirely—to the land-hungry peasantry. Some landlords preferred to switch from feudal sharecropping and debt-labor arrangements to cash rental of their plots—or, in other instances, were pressured by increasingly militant peasants to do so. Finally some peons simply stopped paying rent on their plots as their patrón's control decreased.[17] Thus, by the early 1960s, only some 30 percent of sierra land and less than a fourth of the peasantry were subject to traditional neofeudal arrangements. The remaining land was either rented to the peasantry or was owned by communities and smallholders.[18] In short, in the Andes—with most of Peru's rural population and over 90 percent of its agro-livestock land—the stagnant, unproductive, old order was crumbling before the implementation of agrarian reform.

Land Seizures and Peasant Federations[19]

From 1959 through 1964, peasant unrest exploded across the sierra, set loose by the socioeconomic changes just described. Initially, campesino mobilization centered in two areas: the central Andean mining and ranching state of Pasco; and La Convención, a cocoa, sugar, tea, and coffee-growing valley located on the eastern (high jungle) slope of the Andes. In Pasco, several villages, frustrated by latifundia encroachment on community grazing lands, and angered by conflicts with the Cerro de Pasco mining corporation (which also owned ranches), engaged in a series of hacienda invasions to "recover" land they insisted had been stolen from them over the years. The land seizures (or "recuperations" as the peasants and their supporters insist they should be called) occurred periodically from 1959 to 1962, sometimes resulting in successful occupations, more often leading to bloody police repressions. In La Convención the unrest involved both independent villagers (as in Pasco) and hacienda sharecroppers who formed a union to resist exploitation by their landlords. Led by Hugo Blanco, who came from Cuzco to join them, the campesinos first merely insisted on paying their rent in cash, then refused to pay rent at all, and finally seized the haciendas. Both of the peasant movements led to the formation of *sindicato* (union) federations and elicited support from leftist urban unions and university students.

By 1962, both the Pasco and La Convención movements had been "brought under control," but each planted the seeds of further mobilization in the central and southern highlands. One year later, following the election of reformist President Fernando Belaúnde Terry, a wave of hacienda land invasions swept across the Andes. Belaúnde had actively sought the peasant vote—a significant voting bloc by 1963—and promised land redistribution. His government's initially tolerant attitude toward the seizures led hundreds of peasant communities—primarily in Pasco, Cuzco, and Junín—to occupy hacienda land they claimed was theirs. Ultimately, as many as 300,000 sierra villagers may have taken part in the movement. Land invasions were also accompanied by the formation of peasant federations, often linked with urban radicals—the only political group willing to support the campesino unrest.

In 1964, the Belaúnde administration came under intense pressure from the oligarchy, opposition political parties (including APRA), and the military to stop what many perceived to be a revolutionary threat. Most of the peasant federations were repressed, particularly those in the south that had links with urban Marxists. Some peasants kept their land, however, and Belaúnde's limited agrarian reform program (gutted by the opposition majority in Congress) granted legal sanction to many of the occupations. More important, though the government eventually repressed the peasant federations and crushed a 1965 guerrilla movement, the lesson of the early 1960s unrest was not lost on the military. When reform-oriented officers led by General Velasco ousted President Belaúnde in 1968, they resolved that the only way to bring long-lasting peace to the Peruvian countryside was to enact an agrarian reform which would truly eradicate latifundia domination of land tenancy.

The Military's Rural Revolution

Probably more than any other program initiated by the Velasco government, Agrarian Reform Decree Law 17716 (June 24, 1969) expressed the "radical reformist" modernizing ideology of the military's more progressive sector. Few observers at the time of the reform decree—and certainly not the government itself—foresaw the full political and economic consequences of the proposed change. Critics on the left saw the agrarian reform, and the government's policies in general, as insufficient, bourgeois incrementalist change. Much of the Peruvian upper class and bourgeoisie viewed it (and the government in general) as quasi-communistic and the ruination of the country's productive capacities. In time, it proved to be neither a panacea for the country's ills nor the cause of its economic collapse. Whatever its merits and faults, however, it seriously altered (probably irrevocably) political and economic power relationships in the Peruvian countryside.

The agrarian reform was based on a series of assumptions that were basic to the Velasco regime. Fundamental rural change was viewed as necessary for the following reasons:

(a) social justice demanded that long-standing inequities in the countryside and the domination of the peasantry be ended;

(b) without reform, national political tranquility would always be threatened by the latent potential for rural revolution;

(c) industrialization and broad-ranged economic development could only be accomplished if the rural population was further integrated into the consumer market;

(d) the landed aristocracy was a backward-looking elite whose continued power stood in the way of national industrialization and modernization;

(e) agrarian reform would create rural employment and slow down migration to Lima and other crowded cities;

(f) reform could be used as a means of transferring capital from the insulated, highland agricultural sector into the more modern and dynamic sectors of the economy;

(g) agrarian reform would increase agricultural production (and productivity) thereby better supplying the country's food, raw material, and export needs.[20]

Underlying the reform, then, was the military's anti-oligarchical impulse. Government spokesmen talked openly of "breaking the back of the oligarchy" and, several months after the reform decree was issued, President Velasco declared "the oligarchy is an impediment to a true development of Peruvian industrialization and has always been on the side of international consortiums."[21] To be sure, the military came to office at a time when the power of the traditional landed aristocracy was well into the process of collapse. The new military government was now ready to finish off the tottering hacendado wing of the oligarchy and, at the same time, confront the still-powerful coastal plantation barons.[22] Furthermore, the reform would, it was hoped, have the additional advantage of lining up coastal agricultural workers and the Andean peasantry behind the new regime, no small gain for a government in need of legitimacy.

Within 48 hours of the promulgation of Decree 17716, 8 of Peru's 9 largest sugar complexes were seized (with the ninth expropriated soon afterward). Together these estates had produced 60 percent of the nation's sugar.[23] This sector had specifically been excluded from Belaúnde's 1964 legislation and by expropriating them first (and so quickly) the military clearly demonstrated the seriousness of its intent. Tactically, the swift seizure of these vast plantations denied the rural oligarchy's most powerful sector (with important investments in banking, commerce, and industry) time to organize resistance or to decapitalize. The military also may have seen the expropriations as a way to weaken the *aprista* Sugar Workers Confederation, as APRA was a long-time bête noire of the armed forces.

If the military's motivation was anti-oligarchical, it was not socialist. While the largest haciendas and plantations were to be expropriated, the government envisioned an important role in the new rural order for the middle-sized farmer. Hacendados in the Andes were initially allowed to maintain some 55 hectares of irrigated land (more in some provinces) and twice as much unirrigated. On the coast, landlords could retain up to 150 hectares.[24] Having the right to choose what portion of their property they retained, the hacienda owners obviously kept the most productive land.

While the decision to permit, indeed to encourage, an agrarian "middle class" made sense from a productive and administrative point of view, it failed to anticipate the political problems arising from this group's attempts to limit the reform. The decision also conflicted with the peasantry's and agricultural proletariat's desire to extend land redistribution. In fact, Peru's military rulers initially viewed agrarian reform in highly technocratic terms, seeing it as a means of relocating resources, removing impediments to economic modernization, and increasing production and purchasing power. In conventional military fashion, most government officials envisioned the process of change as being dictated from above, with little room for peasant influence.[25] There was little appreciation of the complex political forces that would be unleashed or of the relationship of land tenancy to the power structure.[26] When the process of land redistribution led to renewed rural mobilization, rather than depoliticization, the armed forces were confused and divided over how to cope.

The 1969-1974 Period

For the first two years after promulgation of the reform decree, government expropriations were confined primarily to the coast. In the sierra, activity was largely limited to Cuzco, Junín, and Pasco—sites of 1960s peasant mobilization—and consisted principally of transferring title to peasant cooperatives of land already held by the agrarian reform institute (under the 1964 legislation) or of property in de facto control of the villagers as the result of invasions. In mid-1971, the Minister of Agriculture announced that the reform had largely been completed on the coast and that future expropriations would henceforth be focused primarily in the highlands.[27]

The process of expropriation is a slow one necessitating inspection of titles, determination of sometimes confusing boundaries, assessment of the land's value, and landlord appeals to an agrarian reform tribunal.[28] Typically, nearly two years elapsed from the time a hacienda was initially marked for expropriation and the time it was actually taken over by the agrarian reform institute (the nine major sugar complexes seized immediately after the reform decree were the major exception). Another year or more might pass until the title to the land was transferred to the peasantry (adjudication).

The complex process of expropriation and adjudication stretched the resources of the Ministry of Agriculture's technicians. Since it seemed impossible to handle the whole country at once, the government concentrated first on the coast, which held over one-fourth of Peru's cropland and 60 percent of its irrigated land in cultivation (see Table 2).[29] The long lapse time, however, allowed many highland hacendados to decapitalize their estates or to divide up title to their land between relatives (though the latter process was far more restricted than in other Latin American nations undergoing agrarian reform). By 1974, the slow rate of reform in the Andes precipitated renewed peasant unrest, which in turn accelerated government expropriations.

The Hacendados Defend Their Interests: 1970-1972[30]

In the face of the "revolutionary" government's resolute attack on the latifundia system, Peru's landed elite found itself in an unaccustomed defensive position. The swift and definitive expropriation of the sugar complexes precluded any

Table 2

Land Use in the Coast and Highlands: 1972

Type of Land	Highlands (Hectares)	Coast (Hectares)
Cultivated	2,280,523	806,198
(Irrigated Crop)	(491,616)	(744,177)
Pasture	14,300,712	495,638
Other (Woods, Mountains)	2,782,226	543,435
Total	19,363,461	1,845,271

Source: *II Censo Nacional Agropecuario-1972* (Lima: Oficina Nacional de Estadística y Censos, 1975). Cited in Fernando Eguren, "La Tierra: Su Distribucion y los Regimenes de Tenencia" (Lima: Desco, 1978).

broad-front attack against agrarian reform. Most of the Peruvian lower and middle classes—in short, the bulk of public opinion—shared the military's anti-oligarchical sentiments. Consequently, landowning interest groups decided to salvage what they could and launched a defense of the "small and medium private property owner." Appealing to the more conservative, technocratically oriented officers within the Ministry of Agriculture and to the interests of the urban middle class, the hacendados tried to limit the damage done them by the reform program.

The landowners had several objectives: to slow down the pace of land expropriation, giving themselves time to organize politically; to decapitalize; to maximize the land ceiling; to maximize opportunities for selling their land at better terms than offered under the agrarian reform or to subdivide it among relatives and friends; and to isolate "radicals" within the armed forces who threatened the sanctity of private property.

The hacendados' political interests were articulated through the National Agrarian Society (SNA), the long-time spokesman of the rural elite, particularly the coastal exporters. The SNA sought to influence public opinion at large as well as attitudes within the military government. From 1970 onward, it launched a public relations campaign aided by editorials in the conservative press. The highly influential *La Prensa* of Lima, owned by Pedro Beltrán, a prominent landowner, was a critical ally in this effort. Ads and editorials stressed the need to respect private property. They also argued that middle-sized haciendas were more efficient than the agrarian reform co-ops and were therefore critical in maintaining food production for the cities (the latter claim designed to elicit urban working-class as well as middle-class support). In a further effort to recast its oligarchical image, in 1969 the SNA reorganized its executive board to include owners of middle-sized haciendas and elected as president Luis Gamarra Otero, an agricultural engineer who owned a 55-hectare coastal hacienda.

SNA's "opening" was more than cosmetic. It was designed to forge an alliance between the old rural elite and medium (or even small) owners of private property. Such a coalition was facilitated by family ties between many large and medium-sized hacendados. Political support also came from the National Society of Industry and from conservative sectors of the National College of Lawyers. At the same time, however, the powerful coastal interests that had long dominated the SNA decided to cast off more traditional Andean hacendados whose political influence was primarily exercised at the local government level (and who were therefore increasingly powerless as the military centralized authority at the national level). Stressing the norms of efficiency, SNA propaganda urged the government to concentrate its efforts on: expropriating the neofeudal highland haciendas; improving the operation of already-expropriated coastal cooperatives; and colonizing new lands in the eastern jungle.

The SNA's energetic defense of middle-sized production units received a sympathetic hearing from Agricultural Minister General Jorge Barandiarán, who declared at a June 1970 press conference that private ownership must be protected since "increases in agricultural production will come through the middle-sized agricultural unit, through professional entrepreneurs with 150 hectares of less."[31] New regulations were issued at that time by the ministry to guarantee the rights of private owners. In March 1970 the ceiling on irrigated crop land was raised from 15 to 30 hectares in 4 Andean provinces.

Thus during the first two years of the agrarian reform the landowners were successful in slowing down the pace of land expropriation. Encouraged by the Agricultural Ministry to divide their land privately, many hacendados—particularly in the sierra where peasants were less well informed and very little expropriation was taking place in the early years—sold portions of their estates to their peons at prices that were more favorable to the owners (and costly to the peasants) than the terms of the reform law. Other landlords took advantage of the government's "private division policy"—clearly designed to encourage subdivision of haciendas among their peons—to transfer title to relatives and circumvent ceilings legislation.

The Peasants React: Coastal Mobilization: 1969-1972

The hacendados' attempts to slow down and contain the sweep of land reform were not entirely successful. Their public relations campaign was challenged both by peasant countermobilization and by the more radical elements of the Peruvian armed forces.

Initial peasant reaction came on the coast, spreading both among those peasants who had failed to get land (or not enough) and within the reform cooperatives, where workers often were disillusioned by the new tenancy arrangements. In October 1969—just months after the promulgation of the land reform decree—and again in early 1970 workers on the "Huando" orange plantation, located in the Chancay valley north of Lima, protested its owners' efforts to evade expropriation by dividing the estate privately. When private subdivision was resumed in mid-1970, a new strike was called (in October) which lasted four months. Featuring several marches on the Presidential Palace, the Huando unrest became a national symbol of the confrontation between hacendado and peasant, and a test of the government's resolve to enforce agrarian reform. The plantation's workers were supported by radical students from the National Agrarian University, leftist urban labor unions, Hugo Blanco (just released from seven years imprisonment) and progressive newspapers such as Lima's *Expreso*. In November 1970 some 10,000 hacienda workers in the coastal Cañete valley struck against private subdivision of latifundia in their region.

If the SNA's attempts to protect the interests of the middle-sized hacendado were supported by conservative officers in the Ministry of Agriculture, protesting coastal peasants were endorsed by more radical military elements, particularly officers in ONDECOOP (National Office of Cooperative Development), the government agency charged with establishing agricultural co-ops. In February 1971—responding to the coastal demonstrations—the government annulled the owner-initiated subdivision of Hacienda Huando and other estates in the Chancay and Huaral valleys.

Expressions of discontent among peasants who had not yet benefited from land redistribution were supplemented by unrest among agricultural workers who *had* been included in the reform process. On the coast most expropriated lands (particularly the sugar plantations) were turned into cooperatives called CAPs (Agricultural Production Cooperatives). Under the terms of the agrarian reform decree, managerial control was to be exercised by a directive board composed of elected representatives from the ranks of the cooperative's workers as well as government-appointed technicians. Although government rhetoric insisted that CAPs would be controlled primarily by the workers themselves ("the land belongs to those who work it"), it soon became apparent that real power often rested in the hands of government technicians ("*técnicos*"). In some instances these new administrators were the same people who had previously worked for the plantations' private owners. While such administrative continuity may have sometimes improved initial efficiency, it added to the workers' feelings that their "masters" had merely changed from private corporations or hacendados to the state. After announcing that the CAPs would be administered by their worker-elected Delegate Assemblies, the government proceeded to designate the majority of delegates on the principal coastal sugar plantations. In the Tumán sugar plantation, 98 of 120 delegates were designated by the state. On the giant Casagrande complex, the government selected 96 of 120. Indeed, in only one of the nation's primary sugar plantations (Cayaltí) did the workers really elect a majority of their Delegate Assembly.[32]

While government control over the critical sugar industry may have been justified on technical grounds or even in terms of the national interest, the contradiction between the agrarian reform's radical rhetoric and governmental behavior reflected both the crippling internal divisions within the armed forces and the military's general failure to grasp the political implications of its own actions. In order to insure the "continuity of technical and administrative direction" on the haciendas (as called for in Article 30 of the Agrarian Reform Law), reform was carried out from the top down ("on behalf of the peasants") with little or no grassroots participation. Even in the election of those delegates to the Delegate Assembly whom the workers did choose, white-collar workers were overrepresented and union officers were barred from holding positions.

Opposition to "external" (government) control of the CAPs was organized by two different groups. In the northern coastal state of La Libertad, members of the centrist APRA sugar workers' union resented the exclusion of their officers from the cooperatives' directorships. More radical resistance to government control came from the Marxist-led plantation workers in Lambayeque. From November 1969 through the end of 1971, strikes and protest demonstrations were carried out on many of the major coastal plantations. Workers on the haciendas "Tumán" and "Cayaltí" were particularly militant and elected CAP representatives who vigorously attacked the government's domination of cooperative administration. In Tumán confrontations between workers and government bureaucrats led to more than 50 arrests.

Initially the government reacted strongly against the coastal unrest. In March 1970 APRA union leaders on the Cartavio plantation were arrested during a strike. As discontent spread, the military issued a decree calling for the imprisonment of those "who block the progress of agrarian reform." The vaguely worded decree was designed to squelch opposition both from the rightist SNA and from militant hacienda workers. Conservative military elements even succeeded in removing some of the government's own ONDECOOP organizers who had proven to be too zealous in their mobilization of the peasantry. In January 1972 hundreds of police occupied the hacienda Tumán and arrested 40 worker leaders.

Gradually government policy began to change as more left-leaning military officers pressed for greater peasant participation in the reform process. One month after jailing the militant union leadership at the Tumán plantation, the government released them and increased the level of rank and file influence in the hacienda's administrative process. This turnabout in state policy was but one of several attempts by the military to respond to the coastal peasantry's unexpectedly vigorous mobilization. These events, coming after the organized campesino protests in the central coastal valleys north and south of Lima from 1969 through 1971 that forced the government to annul owner-initiated parcelization of estates in that region, made clear that a purely technocratic, nonpolitical agrarian reform was impossible. Organized pressure was being exerted on the government from both right and left, forcing a reexamination of basic rural policies.

SINAMOS and the Death of the SNA: 1972-1974

In July 1971, ONDECOOP was replaced by SINAMOS, the "National System of Support for Social Mobilization." Officially designed to serve as a "transmission belt" of demands from the population to the government, SINAMOS was motivated largely by growing rural unrest. Peru's military rulers saw it as an umbrella organization which would stimulate and channel urban and rural mass participation in the revolutionary process and thereby weaken challenges to government authority from the right or the left. For the more radical officers, such as General Leonidas Rodríguez, SINAMOS' first national director, the organization provided a means for increasing peasant participation in the agrarian reform and for combating efforts by the landlords to restrict land redistribution. More conservative military men hoped that SINAMOS would outflank radically led, independent peasant organizations in Lambayeque and Piura.

The creation of SINAMOS intensified hacendado opposition to government policies. Landlords reacted indignantly when radical SINAMOS organizers at the grassroots level not only removed a number of rightist local officials but also, on occasion, supported peasant takeovers of unexpropriated haciendas. An SNA newspaper advertisement in the pages of *La Prensa* indicated the extent of landlord alarm. Entitled "Urgent," it charged that "known professional agitators tied to SINAMOS are entering coastal estates."[33] As the pace of land expropriation accelerated, hacendados hastened to decapitalize their land. Faced with growing landlord intransigence, on May 12, 1972, the military regime issued a decree calling for the creation of a National Agrarian Confederation (CNA) to serve as the sole legally recognized voice of agricultural interests. The most important effect of the new decree was the withdrawal of legal recognition for the hacendado-dominated SNA. In the months that followed, landlords made one more last-ditch attempt to slow the force of rural change. Through the closing months of 1972 and in early 1973, middle-sized owners organized a series of public demonstrations in the coastal state. In June 1973, a march by some 5,000 northern landowners in the state of Piura attacked the "collectivist" nature of the agrarian reform. Three months later the first "National Congress of Small and Medium Farmers and Ranchers" was held in Arequipa. Delegates allegedly representing all the country's 21 states demanded greater landlord input in the agrarian reform, attacked new rules limiting owner-initiated sale and division of estates, and called for the abolition of SINAMOS. As usual, these demonstrations and congresses were accompanied by a public relations campaign in the conservative press.

This time the landlords' efforts attained very limited success. To be sure, in August 1973, conservative voices in the Agriculture Ministry did issue a "Law of Defense of Medium and Small Property" which gave private owners some additional protection. However, the same law also restricted the private purchase of new lands and permitted the expropriation of any middle-sized haciendas whose owners were not providing housing for their workers or were not conforming with national labor legislation (as was commonly true in the rural sector). As more leftist military men secured greater influence in formulating agrarian policy, government spokesmen criticized those who "use the defense of private property [as a means] of attacking the revolution." The July 1974 expropriation of *La Prensa* (and other major Peruvian newspapers) stripped the landowning class of its principal voice. Combined with the earlier abolition of the SNA, this government action helped silence the rural landlord's voice in national politics.

The Growth of the CNA and the Spread of Peasant Unrest

As the pace of land expropriation quickened in 1973-73, and as the government stiffened its position against the rural oligarchy, Peru's military rulers saw a need for greater state control over spiraling peasant unrest. Through SINAMOS and the CNA, policy-makers hoped to direct campesino mobilization against the hacendados while at the same time diverting the peasantry away from more radical, independent organizations. The National Agrarian Confederation (CNA) was designed to include the entire range of agricultural interests. In effect, however, it largely represented the beneficiaries of the agrarian reform and members of the organized peasant communities (*comunidades campesinas* with communal lands). Day laborers, landless peasants, and independent smallholders were only marginally represented.

The CNA was organized through a pyramidal structure with each of the country's provinces having an Agrarian League representing its various base organizations. The leagues would elect delegates to a state-wide federation which would, in turn, elect national representatives for the CNA. The first provincial Agrarian Leagues were established, under the direction of SINAMOS organizers, in October 1972, and four months later the first state federation was created in Piura (a center of peasant unrest at that time). Although several additional federations were organized in 1973, it was not until September 1974 that a sufficiently extensive rural network was created to justify holding the CNA's first national congress.

Peruvian sociologist Mariano Valderrama maintains that local base organizations or even provincial Agrarian Leagues were often initiated by SINAMOS organizers (called "promoters") with little or no peasant involvement. In one province, only three of the league's nine base organizations ever actually held meetings. In other instances, local CNA officials were elected who were government bureaucrats having little contact with the peasants they allegedly represented. Finally, the CNA was mistrusted by more militant campesinos who often viewed it as little more than the government's arm for controlling the peasantry. Marginal peasants—the landless and those with holdings too small to be visible—initially perceived the CNA as the voice of richer campesinos, particularly those who had benefited from the agrarian reform.

Given these early difficulties, it is small wonder that the government-sponsored confederation did little to stem the rising tide of rural discontent in various parts of the country. By 1973, the magnitude of rural discontent had grown both in terms of geographical spread and the intensity of protest. During 1974 peasant militancy reached a climax in the highland province of Andahuaylas, located in the south-central state of Apurimac. Inhabiting one of the most feudalistic, latifundia-dominated regions of Peru, the campesinos of Andahuaylas had greeted the 1969 Agrarian Reform Decree with great hope and expectation. As one local leader put it, "When we heard on the radio that they had thrown out the owners of the great coastal haciendas, we said '[President] Velasco is a man!' We thought that here in Andahuaylas they would do the same, that they would throw out the rural bosses [*gamonales*] who had robbed our land, those who had killed our animals when they entered the pasture." [34] Yet, more than five years later, Andahuaylas remained virtually untouched by the reform. As of June 1973, only 4 of the province's 130 haciendas had been expropriated. Peasants watched with mounting frustration as local hacendados decapitalized their estates, selling off animals and equipment in anticipation of eventual expropriation. Finally, on July 15, 1974, militant campesinos, organized into FEPCA (the Provincial Federation of Andahuaylas Peasants), seized a local hacienda. Other land invasions soon followed and during the next three months peasant unrest spread across the province until a total of 68 haciendas had been seized. Working closely with Vanguardia Revolucionaria, FEPCA leaders coupled their land seizure movement with attacks on "capitalist imperialism and the bourgeois state."

Surprised by the intensity of the Andahuaylas movement, the government initially took a conciliatory position, entering into negotiations with FEPCA leaders and ignoring hacendado calls for the expulsion of peasants from the invaded estates. At the same time, however, government-controlled newspapers as well as the newly organized CNA condemned the wave of invasions and called on peasants to obey the law and not "sabotage the agrarian reform." Weakened by its rather utopian demands for sweeping national change and by intense repression including the arrest of critical leaders and rank and file, FEPCA soon collapsed as an independent organization. Its more moderate leadership was absorbed into the CNA. Most of the invaded haciendas, however, remained in the hands of the peasantry. More important, the pressures exerted by the militant peasants in Piura, Andahuaylas, and other mobilized regions of the sierra led the government to carry forward land expropriations at a higher rate.

Land Redistribution 1974-1979

In many ways 1974 marked the apex of the radicalization of Peru's land reform. The Andahuaylas and Piura grassroots mobilizations, the reorganization of the Confederación de Campesinos del Perú (CCP), the first national congress of the CNA, the increased militancy of SINAMOS, and the expropriation of the conservative press—all these events coming on the heels of the dissolution of the SNA—shifted the balance of rural political and economic power leftward. It was a shift that was to be shortlived since the following year witnessed the overthrow of the Velasco government by more conservative military officers led by General Francisco Morales Bermúdez. The events of 1974 had nevertheless created an inexorable momentum toward the completion of land redistribution, a momentum the Morales government did not try to reverse. Table 3 reveals the scope and pace of reform through mid-1979.

Table 3

Rate of Land Expropriation and Adjudication[35]

Date	No. of Beneficiaries (families)	Expropriated Land (hectares)	Adjudicated Land (hectares)
June 1971	90,000 (approx.)	1,600,000 (approx.)	n.d.
December 1974	200,000 (approx.)	n.d.	4,181,000
July 1976	279,600	8,066,000	6,810,000
December 1978	347,000	8,932,000	8,070,000
June 1979	370,000 (approx.)	n.d.	9,000,000 (approx.)
Anticipated Final Total	400,000	9,500,000	9,750,000

Source: Dirección General de Reforma Agraria y Asentamiento Rural.

Thus, half the 21 million hectares of agro-grazing land in the coast and highlands have been expropriated from the hacendados and turned over to the peasantry. Another 3 million hectares (15%) was already in the hands of small farmers (under 20 hectares) before the reform. Officially recognized peasant communities, which controlled about 6.6 million hectares before the reform, received nearly 900,000 more from the redistribution process. Less than 10 percent of all land—1.5 to 2.0 million hectares—was left in medium-sized haciendas (20-150 hectares) and none in the huge latifundia that once dominated Peruvian land tenancy.[36]

The bulk of the expropriated coastal estates, as noted, were turned into cooperatives, called CAPs, where the land was supposed to be utilized as a single production unit and CAP members were to be paid workers (as most had been on the more modern plantations before the reform). The beneficiaries of the agrarian reform within the CAPs were thus the estate's workers and not neighboring villagers or independent smallholders who may have hoped to gain hacienda land from the redistribution.

In the sierra, CAPs were also formed but the bulk of expropriated land was converted into units called SAIS (Agricultural Society for Social Interest). The SAIS sought to meet the often-conflicting demands for land of the former hacienda peons and the villagers of surrounding communities. As many as 30 nearby peasant communities might be joined with the peons of one or more haciendas to form a SAIS. Villagers would vote for the directive board and could sit on it (indeed, in many SAIS they had a voting majority). They would also share in any profits the cooperative might make beyond workers' salaries, operating expenses, and reinvestment. However, only the haciendas' former feudatarios actually worked for the SAIS and got use of its land (overwhelmingly pasture).

The purpose of creating CAPs and SAIS was to preserve the haciendas as productive units and not break them up into smallholdings, which government planners felt would lower agricultural output. Original planning called for turning 80 percent of agrarian reform land into CAPs and SAIS, with the remaining 20 percent to be given to peasant communities, "peasant groups" (groups of smallholders who banded together into a cooperative in order to receive agrarian reform lands) or, in a small number of cases, to individual smallholders. Ultimately, the proportion going to the CAPs and SAIS has been somewhat lower than planned.

Table 4 clearly shows the government's preference for maintaining large production units operated as cooperatives of some kind. Only in exceptional cases is land transferred to individual smallholders. The 60 vast highland SAIS each had an average of 1,000 families (including *comuneros* and over 46,000 hectares of land (largely pasture).

In actuality, the degree of cooperative farming has been far less than the data in Table 4 suggest. Economist José María Caballero notes that most of the SAIS (and some CAPs), while technically classified as cooperatives, in fact have continued

Table 4

Actual Adjudication of Agrarian Lands: 1967-1978 [37]

Type of Unit	No.	No. of Benefiting Families (%)	Area of Land Received (%) (hectares)
CAP	515	103,806 (29.9%)	2,264,411 (28.1%)
SAIS	60	60,930 (17.6%)	2,775,757 (34.4%)
Peasant Communities	427	114,511 (33.0%)	856,232 (10.6%
Peasant Groups	747	44,043 (12.3%)	1,606,612 (19.9%)
Social Property	8	1,375 (0.4%)	223,000 (2.8%)
State Property			123,372 (1.5%)
Individuals		22,301 (6.4%)	220,932 (2.7%)
Total	1,757	346,966	8,070,316

Source: Dirección General de Reforma Agraria y Asentamiento Rural.

to farm their land as individual lots. This is particularly true in the highlands where the SAIS peasants seem committed to having their own land. On the coast—particularly on the sugar plantations—workers are accustomed to working for a large enterprise, realize that it would be absurd to break up their highly mechanized estates, and know full well that the government would not permit subdivision even if the CAP members preferred it. In all, about two-thirds of SAIS and CAP lands (particularly the highland SAIS) are actually composed of individual plots. Of the land that is really run cooperatively as a single unit, 85 percent is on the coast.[38]

As of the end of 1978, the redistribution of land was nearly complete. Ninety-four percent of the 9.5 million hectares scheduled for expropriation had actually been expropriated; 83 percent of the final total had been adjudicated to the peasantry; and 87 percent of the planned beneficiaries had received their land (Table 3). The Peruvian 1969 Agrarian Reform decree was but one of several issued in the 1960s throughout Latin America as a response to the Cuban Revolution and the perception by international agencies, the Alliance for Progress, and many Latin American officials that change was needed. Venezuela, Colombia, Ecuador, and Chile were other nations with much-touted programs (Mexico and Bolivia had experienced earlier redistribution as a result of their revolutions). Of these, only Peru actually eradicated the latifundia system.

Peasant Mobilization on the Eve of the '80s

During the early 1960s, when thousands of Peruvian peasants were challenging the old order, few would have predicted that the national government would sweep away that order in the near future. Fewer still would have anticipated the renewed level of campesino discontent that has recently manifested itself in spite of, or because of, the agrarian reform.

Perhaps the clearest indication of the government's failure to marshal peasant support has been the growing radicalization of the Confederación Nacional Agraria. Created by the military as a counterbalance to independent Marxist peasant organizations, the CNA has increasingly opposed its government sponsors. From its very inception, the confederation included provincial and department leaders—from states such as Lima, La Libertad, Ancash, and Cuzco—who wished to maintain their distance from government decision-makers and to act as an independent pressure group on rural policy. Initially the CNA tended to represent the CAP, SAIS, and peasant communities who had directly benefited from the agrarian reform, while the radical CCP was more representative of campesinos who had been left out of the redistributive process. Gradually, however, that distinction has blurred and the two organizations have moved ever closer to each other in political outlook.[39]

As early as November 1974, the more militant CNA Agrarian Leagues in the states of Lima and Piura were supporting invasions of unexpropriated haciendas. In other parts of the country, local CCP and CNA leaders coordinated protest marches and occasional land seizures. By mid-1976 (after a brief period of intensified land redistribution and efforts by the new Morales

Bermúdez government to woo militant CNA leaders), relations between the military regime and the confederation had deteriorated sharply. The widening split was manifested in July of that year by the brief arrest of the CNA's Secretary General for the state of Lima.

Early in 1977, Avelino Mar, a militant peasant leader from the Convención Valley of Cuzco, was elected national Secretary General of the CNA. The election of a leader identified with the most radical wing of the Velasco regime reflected the victory of the CNA's militant wing over local officers wed to the government bureaucracy. Charging that the Morales Bermúdez government had abandoned the agrarian reform, Mar and other CNA national officers called for reductions in the amount of land allowed middle-sized haciendas, more extensive expropriations, and the creation of a government bank to assist the low income agricultural producer.

At a December 1977 CNA Assembly, the organization's militants strengthened their position by voting to expel government-supported state leaders in Junín, Pasco, and Huánuco. Shortly thereafter, the Minister of Interior, General Luis Cisneros Vizquerra, charged top CNA officers with conspiring against the government. Six months later, following a CNA-endorsed nationwide general strike, the government announced the dissolution of the Agrarian Confederation at the national level.[40] Departmental federations and provincial Agrarian Leagues could continue to operate, but the confederation's national organization was denied legal recognition. At the same time, Lima's *El Comercio* and other government-controlled newspapers launched a press campaign against alleged "subversion" by CNA officers. Secretary General Avelino Mar was forced to go into hiding after he was accused of conspiring to halt food shipments into Lima during the general strike.[41]

By the close of 1978, the CNA and the CCP were reportedly on the verge of a merger. While both organizations were technically illegal at their national levels, they continued to operate rather openly. In an interview with an officer of the CNA's departmental federation in Cuzco (FARTAC), I was told that the policy differences between the CNA and CCP were minimal. On the wall of the FARTAC headquarters, posters advocating socialism and endorsing the Cuban Revolution suggested the extent of the federation's radicalization.

The frustration and anger of nonbenefiting peasants toward the agrarian reform is demonstrated by yet another recent phenomenon. Between 1973 and 1975 there were a series of peasant land seizures principally directed against unexpropriated haciendas. In 1975, however, several peasant communities invaded nearby agrarian reform cooperatives from which they had been excluded. During the next four years similar invasions of CAPs and SAIS spread to Ancash, Cajamarca, Cuzco, Huancavelica, Lambayeque, and Piura.

Within the coastal sugar CAPs, deteriorating real incomes and tightened government control have renewed tensions between workers and government administrators. In 1977, after five years of relative labor tranquility, an extensive strike of sugar workers took place. Labor unrest on the sugar estates has further intensified in 1978-79.

The Roots of Discontent

Why has the military government's agrarian reform program, so far-reaching in its redistribution of latifundia land, been so unsuccessful in satisfying much of the peasantry? Some of the underlying difficulties can be attributed to the manner in which the reform program was conducted. Other problems have arisen from broader government economic policies that have undercut the reform's benefits. Finally, certain difficulties either are beyond the government's capacities to control or cannot be solved merely by agrarian reform.

Peasants vs. Bureaucrats. Though the Peruvian peasantry initially greeted the government's agrarian reform decree with anticipation and hope, they were less enthusiastic about the arrival of the first state bureaucrats. In the Andean highlands particularly, the sociocultural gap between local Agriculture Ministry officials and the peasants whom they allegedly serve has always been enormous. The white or mestizo bureaucrat, wishing he could have been stationed in a comfortable office in Lima and highly resentful of being assigned to the hinterlands, often treats the Indian peasantry with contempt. One Scottish anthropologist, who had developed a close relationship with the peasants of a Cuzco cooperative, told me of seeing a CAP bureaucrat physically abuse a co-op member of having become "too friendly with the gringo."[42] Those government bureaucrats who had more positive and progressive orientations (as in the case of many ONDECOOP and SINAMOS organizers) often found themselves falling victim to the campesinos' understandable distrust and suspicion of all outsiders. In some sierra CAPs and SAIS (including the Cuzco cooperative just discussed) the peasants physically expelled government administrators from their land.

On the modernized coastal plantations there is obviously a far narrower cultural gap between the bureaucrats and workers. However, given the importance of the sugar, cotton, and rice estates to the nation's economy, the government has felt a stronger need to make basic economic decisions in the administration of these cooperatives. That intention has brought the state into conflict with the militant trade unionism of many coastal agricultural workers. During the early 1970s, when participatory values had more currency than they now do within the government (particularly when sugar prices were booming), concessions could be made to worker demands. Since the collapse of sugar prices and the concurrent decline of the Peruvian economy, tensions have risen sharply. Despite a 1976 government decree outlawing strikes on the sugar cooperatives, there have been several work stoppages in recent years.

The Shortage of Agrarian Reform Lands. If their grievances against government bureaucracies gave CAP and SAIS workers grounds for complaint, how much more has been the frustration of the many peasants who have failed to receive any land at all. Government statistics indicate that some 400,000 rural families will ultimately share in agrarian reform benefits. While estimates of the size of the agricultural work force vary considerably, the government figures suggest that only 25-33 percent of the agrarian population will benefit directly from land redistribution.[43] Moreover, José María Caballero points out that government estimates of the number of direct beneficiaries are exaggerated both because many families are not receiving enough land to satisfy the government's own minimum subsistence criteria and because many of the 61,000 families listed as SAIS beneficiaries (Table 4) are members (comuneros) of organized peasant communities (comunidades campesinas). They share in SAIS decision-making and in any of the enterprise's profits, but receive no land. Since divisable annual profits (after all expenses, including salaries, are paid) at one of the nation's largest and richest SAIS (Cahuide) only totaled $30 (S/1,324) per family in 1972-73, the economic benefits of SAIS membership for the comunero beneficiaries seem rather dubious.[44] Caballero estimates that only 45 percent of the peasant families which needed land will receive it.

To some extent the problem is beyond the Peruvian government's capacity to solve. With only 0.18 hectares of crop land per person in the country today (a density comparable to India's), Peru has one of the most unfavorable man/land ratios in the world. Fernand Eguren calculates that if all the nation's estates with over 50 hectares of cropland or over 500 hectares of pasture were divided equally among the approximately one million peasant families with holdings of under 5 hectares (including the landless), each family would receive only 0.41 hectares of irrigated cropland, 0.36 hectares of dry cultivable land, and under 6 hectares of pasture.[45] In short, even if the demands of the CCP to expropriate the middle-sized hacienda were accepted by the government, it would not nearly solve the land shortage. Caballero estimates that if the maximum amount allowed any private owner were only 5 hectares (i.e., if all but smallholders were eliminated), the pool of land available for the agrarian reform would increase by only 15-25 percent.

Experts do feel that, with extensive introduction of irrigation, better soil management techniques, and reasonable colonization of the eastern jungle, the amount of agricultural and grazing land in Peru could be increased by some 60 percent. Yet, a combination of urban expansion onto farming lands, the neglect of hacienda irrigation systems as their owners anticipated expropriation, and serious drought in 1977-1979 have slightly reduced food-producing lands. Considerable investments were made in new irrigation projects (particularly before the 1975 bankruptcy of the economy), but most of the payoff lies in the future and many projects are still in the pipeline. Moreover, most of the anticipated irrigation projects, unfortunately, are devoted to increasing sugar and corn cultivation rather than to the production of needed foods.

It is one of the greater contradictions of the agrarian reform that a program designed to defuse latent rural unrest has largely excluded and frustrated the most potentially explosive sector of the peasantry, the very sector that helped force the reform. During the massive peasant mobilization of the early 1960s—the land invasions and campesino sindicato demonstrations—it was the independent comunidades that formed the core of the movement, while the hacienda peons remained more isolated and unorganized. The military government's 1969 reform decree was, in part, a belated response to that unrest.[46] Yet, once the decree was issued, those political factors were quickly forgotten and replaced by more technical and economic norms. Not wishing to divide up the latifundia, and believing that agricultural output would decline if neighboring communities were afforded the use of cooperative lands, military planners turned the haciendas over to their workers and peons, thereby effectively excluding the comunidades from benefits. Peasant communities gained only marginally useful participation in the SAIS and limited land grants (about 11% of the area redistributed—Table 4). Indeed, many communities whose lands had been encroached upon over the years by neighboring haciendas, and which had been engaged in long legal battles to establish their village titles, found their "property" included in the new agrarian reform cooperatives. It should not be surprising, then, that since 1976 there has been a rising number of comunidad invasions of co-op lands.

Cities Against the Countryside. Yet another irony of the Peruvian government's rural policies is that, during the very period in which the regime has tried (with only limited success) to satisfy the peasantry's most basic demand by giving them land, it has simultaneously implemented agricultural policies that are highly prejudicial to the countryside. Under both the Velasco and Morales Bèrmúdez governments, economic policy has favored urban interests over rural ones.

The level of technical aid offered the agricultural producer today is proportionately lower than it was before the agrarian reform. During most of the 1970s, the Agriculture Ministry's personnel and economic resources were so tied up in the process of land redistribution that aid for production was reduced. Now, with the redistributive process complete, the nation's severe economic crisis makes increased state investment in the countryside impossible.

Government import and price policies have also been fairly consistently biased against the agricultural producer. The Peruvian state controls the amount of food imported, the prices paid agricultural producers for many crops, and the cost of many inputs. Consequently, the state largely determines the economic context of agricultural production. As runaway inflation has vastly escalated production costs, the government has kept down the prices paid Peruvian farmers in order to mollify the politically critical (and volatile) urban population. Until 1979, crop prices have been set consistently below world market levels, and have sometimes even failed to meet production costs. As economist Richard Webb notes:

> *Price policy is discriminatory; food imports continue to be exempted from tariffs while the average duty on all other goods has risen from about 30 percent in 1961 to over 70 percent in 1970. Cheap beef imports compete with what is often the principal source of cash income for small farmers.... Tariff-exempt wheat imports favor the consumption of bread; a nondiscriminatory general tariff, or a reverse discrimination in favor of domestic foods, would raise Sierra income from potatoes and other substitutes for bread.*[47]

In short, the common Third World imbalance between city and countryside has been maintained in the Peruvian "revolution." Political scientist Henry Pease García notes that prior to the agrarian reform the rural oligarchy—particularly coastal agribusiness—was a very powerful interest group that could pressure the national government for loans, technical assistance, and more favorable import policies. Now, with the destruction of the landed elite as a power contender, no agrarian political force has developed to take its place. The peasantry has been too preoccupied with the issue of land redistribution and too weak organizationally to serve as an effective pressure group. Thus, the political power of the agricultural sector has actually deteriorated as a result of the agrarian reform.

The National Economic Crisis. To some extent, the current malaise of many Peruvian peasants and agricultural workers is more closely attributable to the nation's economic crisis than to defects inherent in the agrarian reform. During the early years of the military revolution, the government borrowed very heavily from international agencies and foreign banks in anticipation of revenues from new oil explorations and expected price rises for the country's copper and sugar exports. Since 1974, the economy has been driven to the point of bankruptcy by the drastic decline in sugar prices, falling copper revenues, a sharp drop in the fishmeal catch, and the failure of jungle oil fields to yield hoped-for production.[48] Poorly chosen, high capital investments, excessive arms purchases, and administrative mismanagement compounded the crisis. Consequently, the past three years have been marked by severe depreciation of the national currency (from approximately 62 soles to the dollar to 220), enforced budgetary and import austerity, and drastically declining real incomes.

For the sugar workers, dramatic increases in the world price of sugar in the years following their estates' expropriation had led to sharply increased incomes. After 1974, however, the price of sugar fell precipitously and the real income of CAP workers (like that of the entire Peruvian working class) fell some 40 percent. The economic crisis caused the government to roll back the gains in self-management autonomy the workers won through their 1972 demonstrations and in 1976 to issue a decree denying sugar workers the right to strike. The 1979-80 recovery in the world sugar market is too recent to have had an impact on workers' incomes or to predict patterns for the future.

Estimates of the sierra peasantry's living standards are far more uncertain and difficult to calculate. Various case studies of highland SAIS and cooperatives during the early period of land redistribution seemed to indicate that the reform was bringing most beneficiaries rising living standards as they were freed of their former rent obligations. However, rampant inflation in recent years and the failure of crop prices to keep up with the peasants' costs have undoubtedly eradicated all or most of those gains. Indeed, to the extent that the agrarian reform has integrated peasants into the commercial market, they have been more vulnerable than previously to downturns in the national economy. Of course, for the smallholders and the "occasional" (nonpermanent) SAIS and CAP workers who have not shared in agrarian reform benefits, the picture has been even more negative.

The government's budgetary squeeze has also infringed upon rural development, technical assistance, and loan projects for the peasantry. Moreover, the crisis has moved the state to assert greater control over the cooperatives and to centralize the decision-making process even further. Such developments have undermined the desires of peasants and workers (and earlier government promises) for greater self-management of the cooperatives.

Agrarian Reform and Agricultural Production: Food for the Cities

Clearly the military regime's agrarian reform—far-reaching though it was—has failed to win the national government full loyalty of the peasantry, or to co-opt campesino mobilization. How successful has the program been in feeding the nation, particularly the rapidly growing urban sector? An underlying assumption of the 1969 decree was that a large portion of the country's hacendados, especially in the highlands, were wedded to antiquated agricultural techniques and labor relations. Now, motivated by their new control over their own land and aided by government's technically oriented bureaucrats, the peasantry are expected to produce more on the reform cooperatives. Yet, there seems to have been little improvement in land use or in agricultural technology. If anything, disinvestment by landlords facing expropriation has reduced rural productivity. Total output for basic foods and key export crops has remained stagnant or fallen in the past decade.

Tables 5 and 6 reveal that, since the base years of 1961-1965, absolute production of beef, wheat, cassava, and cotton has fallen and there has been a marked steady decline in overall per capita agricultural and food production. Of course, the indices in Table 6 include a sharp agricultural decline from the base years through 1968, predating the military regime's agrarian reform. If recent production data are compared with 1968 (the year before the reform decree), one finds an absolute increase of 9 percent in total agricultural production and a 16 percent rise in total food output (Table 6).

Even this more optimistic interpretation of the data shows that the modest improvement in food production over the past decade has failed to match population growth. Overall agricultural output rose by 2.4 percent in both 1973 and 1974, but increased by only 1.0 percent in 1975. After a 3.3 percent increase in 1976, total output failed to grow at all in 1977 and rose only 0.5 percent in 1978.[49] Absolute production of beef and all basic crops has fallen since 1974 (Table 5). During the first decade of the agrarian reform, per capita food output declined approximately 9 percent.

Critics of the agrarian reform are quick to blame this poor agricultural record on the redistributive process. Marking the tenth anniversary of the reform, a Lima newspaper editorial claimed that the agrarian sector has been ruined by incompetent government bureaucrats managing the CAPs and SAIS. Private owners, said the conservative newspaper, could produce far more efficiently. Looking for another scapegoat, a government bureaucrat told me that when he was assigned to a coastal cooperative soon after its expropriation, he couldn't get the workers to exert themselves because they insisted that they were now their own bosses and could do as they pleased. Still other critics charge that the highland peasantry are not sufficiently educated to manage their own land.

Table 5

Total Production of Basic Commodities: Selected Years, 1961-1978 (000 Metric Tons)

Year	Corn	Potatoes	Cassava	Rice	Wheat	Sugar	Cotton	Beef	Dairy
1961-65 (aver.)	490	1,487	436	324	150	7,373	140	91	
1968	533	1,592	399	286	119	7,226	101	90	
1970	615	1,929	498	587	125	7,530	91	85	
1972	589	1,750	490	552	140	8,582	73	96	
1974	600	1,722	469	426	150	9,179	88	95	634 (1975)
1976	700	1,670	402	579	148	8,761	57	84	640
1977	720	1,600	410	547	130	8,825	58	85	630
1978 (est.)	550	1,650	410	400	90	8,400		83	635

Source: *U.S. Embassy, Agricultural Attaché, Annual Agriculture Situation Reports, 1976-78; U.S. Department of Agriculture, Indices of Agriculture in the Western Hemisphere, 1968-77.* Both sources drew data from the Peruvian Ministry of Agriculture.

Table 6

Indices of Crop, Agriculture and Food Production: 1968-1972 (1961-65 = 100)

	1961-65	1968	1970	1972	1974	1976	1977 (prelim.)
Crops	100	93	108	102	102	100	100
Total Agriculture	100	94	105	101	104	101	102
Total Food	100	98	116	113	115	116	116
Population	100	115	122	130	138	146	151
Per Capita Agric.	100	81	86	78	75	69	68
Per Capita Food	100	85	94	87	83	80	77

Source: U.S. Department of Agriculture, *Indices of Agricultural Production in the Western Hemisphere: 1968-77* (Washington: 1978).

More careful analysis of the types of agricultural and grazing lands that have been directly affected by the redistributive process suggests that output declines have been no sharper on the CAPs and SAIS than on land that has remained in private hands. Indeed, most food produced for the cities has long been grown on units of land too small to have been expropriated. Ministry of Agriculture estimates indicate that 90 percent of the nation's potato production, 84 percent of the corn, 90 percent of the wheat, and 95 percent of onion production comes from units that fall below the law's ceilings. Indeed, Fernando Eguren points out that most of the sierra's latifundia were devoted primarily to ranching or dairying while the coastal estates were mostly in sugar.[50] In short, over 80 percent of the nation's CAP and SAIS lands are composed either of pasture or of sugar lands. Thus, it seems far-fetched to blame declines in major food crops on mismanagement of the cooperatives or on peasant laziness. To be sure, many medium and small farmers may have reduced their investment level fearing that the government would lower the ceilings and expropriate their property. With the completion of expropriations in 1979, that fear should abate.

Undoubtedly, the redistributive process has caused dislocations that reduced production. Declines in productivity, however, can be attributed more directly to the decapitalization of farms and the neglect of irrigation systems than to the failures of the cooperatives. On the sugar plantations, where immediate expropriation precluded decapitalization or sabotage by the owners, production rose sharply from 1968 to 1974 (some 27%—Table 5) and has fallen only slightly since the 1974 collapse of sugar prices. Expert observers of rural Peru note that many cooperatives started off their existence with virtually none of the necessary farm machinery, trucks, irrigation pumps, or even office equipment because these items had either been sold by their owners, been moved onto the land that the hacendado was allowed to retain, been allowed to run down, or even been sabotaged by bitter landlords. Given these adverse conditions and the peasantry's lack of prior opportunity in management of large landholdings, the production record has not been that bad.

Obviously, many peasants, deprived of educational opportunities in the past, could benefit from government-sponsored training programs and technical assistance. Such programs, however, have been as woefully insufficient as has the general level of loans and economic assistance. The country's current economic crisis and need for austerity offer little hope that rural assistance will be increased, or even maintained at its present inadequate level.

Finally, serious drought in the northern part of the country and some central regions has been a critical factor in the nation's recent poor agricultural record. For example, water levels in Lambayeque's Tinajones reservoir began to fall seriously in 1975. By early 1978, the reservoir, which once held 300 million cubic meters of water, was virtually dry, cutting off critical irrigation for the state's rice plantations. Since 1976 drought has cut production of potatoes, corn, rice, sugar, and wheat.

Conclusions

Despite the obvious difficulties and inadequacies of the agrarian reform, it would be wrong to categorize Peru's rural experiment as a failure. If the redistribution of land has not led to obvious improvements in the peasants' living standards or reductions in rural-urban migration, it is likely that the situation would be worse without the reform. Perhaps the most important contribution of the program is that it has more rapidly terminated the already declining system of restraints which the old order placed on the hacienda peasantry and their village compatriots. On many haciendas, peasants who formerly were subject to the whims of the local landowning bosses now are directing their own cooperatives. Even where government bureaucrats have replaced the hacendados as "the new masters," greater opportunities exist for workers and peasants. The agrarian reform has integrated the peasantry more fully into the national economic and political systems and has opened the door for peasants and agricultural workers to assert some control over their own destinies. Initially, the reform unleashed dormant peasant unrest, but without it, peasant discontent (and the resulting political instability) would have been far greater during the Peruvian economic collapse of 1975-1979.

What has become abundantly clear in the past decade, however, is that land redistribution in itself is insufficient either to solve the problem of rural poverty or to produce more food. At the very least, the government must help increase the area cultivated through fuller use of already existing crop and pasture lands, increased irrigation, colonization of new areas, and better land management. Agricultural technology must also be improved. From the nutritional perspective, where it is ecologically possible (given altitude and other natural considerations) some of the sierra's extensive pasture should be converted to crop cultivation. State pricing policies should be altered to increase the profitability of raising nutritionally rich crops and dairy products and to reduce the cash incentives for producing meat.

On a more fundamental level, the countryside continues to receive an inadequate share, relative to the urban sector, of government resources and of national income. In Peru, as in most of the Third World, there is a vast income distribution inequity between city and countryside. Typically, urban per capita incomes are some four times greater than in rural areas. Richard Webb notes that from 1960-1970, under the Belaúnde and early Velasco governments, a certain degree of downward income redistribution took place within the urban and rural sectors, but virtually none occurred between sectors.[51] Through the 1970s the government has failed to correct that imbalance. As we have seen, state agricultural pricing policy has a strong urban bias.

Increasing the prices paid to Peru's agricultural producers, and the accompanying jumps in food costs to the consumer, would undoubtedly alienate the urban middle and working classes. Indeed, at the present time, with annual inflation rates of some 70 percent and a 40 percent decline in workers' real incomes over the past five years, price rises would soon precipitate urban rioting and strikes. Consequently, peasants can anticipate no significant improvements in the near future. Similarly, the austerity measures imposed on the Peruvian government by international lending agencies (particularly the IMF) currently preclude increased government loans and technical assistance to the cooperatives and small farmers.

Ultimately, however, higher agricultural prices would serve the interests of the cities as well as the peasantry by stimulating greater food production. Moreover, until government policies are introduced which redistribute income and resources to the countryside, the nation can continue to expect ongoing peasant discontent and unrest.

(August 1980)

NOTES

1. After seizing power in October 1968, the Peruvian military government instituted a series of left-nationalist reforms—including its far-reaching agrarian reform decree—not normally associated with the military regimes in Latin America. Since a 1975 internal coup, however, the government has moved to the right. In the wake of economic collapse and the absence of popular support, the armed forces relinquished its control to a civilian government in early 1980. In June 1978, as a first step toward that transition, a national Constituent Assembly was elected. For analyses of the "military revolution" see: Abraham Lowenthal (ed.), *The Peruvian Experiment: Continuity and Change under Military Rule* (Princeton, N.J.: Princeton University Press, 1975); Henry Pease García, *El Ocaso del Poder Oligárquico* (Lima: Desco, 1977). The latter contains an exhaustive 1,000 item bibliography. On the transition process, see: Howard Handelman and Thomas G. Sanders, *Military Government and the Movement Toward Democracy* (Bloomington, Ind.: Indiana University Press, 1981).

2. Blanco helped lead a peasant land seizure movement in Cuzco's high jungle region in the early 1960s, a movement which later contributed to peasant mobilization throughout the highlands. A legendary figure among the Quechua peasantry, Blanco has more recently amassed a great personal following within Lima's shantytowns and received the third highest individual vote of any candidate in the June 1978 Constituent elections. See: Hugo Blanco, *Tierra o Muerte* (México: Siglo XXI, 1972).

3. *El Comercio* (Lima: June 25, 1969).

4. The classic English-language work on Peruvian agriculture and land tenancy as of 1960 is Thomas Ford, *Man and Land in Peru* (Gainesville, Fla.: University of Florida Press, 1962); socioeconomic structure in the rural sierra is discussed in Howard Handelman, *Struggle in the Andes: Peasant Political Mobilization in Peru* (Austin, Tex.: University of Texas Press, 1975).

5. During the past two decades sugar, fishmeal, and minerals (principally copper) have been Peru's major exports. Currently, mining occupies the first rank with coffee, having joined the leading group in export value.

6. Liisa North, "Political Conjunctures, Military Government and Agrarian Reform in Peru" (unpublished paper, 1978). Some sugar plantations are foreign-owned (Grace Corporation, etc.), but most belong to the Peruvian National Oligarchy.

7. On APRA's reformist ideology and origins, see: Peter Klaren, *Origins of the Peruvian Aprista Party, 1870-1932* (Austin, Tex.: University of Texas Press, 1973); Liisa North, "Origenes y crecimiento del partido aprista peruano" *Desarrollo Económico* 10 (No. 38), 1970. Though long Peru's single largest party, APRA has never been able to take power.

8. Julio Cotler and F. Portocarrero, *Organizaciones campesinas del Perú* (Lima: Instituto de Estudios Peruanos, 1967). I have adjusted and rounded off the figures to compensate for changes in the past 12 years.

9. Roberto MacLean y Estenos, *La Reforma Agraria en el Perú* (México: U.N.A.M., 1959), p. 40; Hugo Neira, *Los Andes: Tierra o Muerte* (Madrid: Editorial ZYX, 1968), pp. 41-44.

10. The use of paternalism, mixed with repressive controls, as a means of containing peasant unrest is discussed in James Scott, *The Moral Economy of the Peasant* (New Haven, Conn.: Yale University Press, 1975). On Peru, see Handelman, *Struggle;* François Bourricaud, *Cambios en Puno* (México: Instituto Indigenista, 1967).

11. For a discussion of hacienda mobilization: Julio Cotler, "Traditional Haciendas and Communities in a Context of Political Mobilization in Peru" in R. Stavenhagen, ed., *Agrarian Problems and Peasant Movements in Latin America* (New York: Doubleday, 1970).

12. Peru in the 1960s had some 1,500 officially registered peasant communities (with communally owned lands) and over 4,500 unrecognized ones. See: Roberto MacLean y Estenos, *Sociología del Perú* (México: U.N.A.M., 1959), p. 262; Handelman, *Struggle*, pp. 28-36.

13. Some 20 percent of all peasants had some feudal obligations, of whom perhaps one-third were full-time hacienda peons and the remainder, smallholders who worked for haciendas. Diego García-Sayán "Estudio de Cambio Rural." (Unpublished prospectus for a research project for Desco, Lima, 1978).

14. Jean Piel, "A propos d'un soulèvement rural péruvien au début du vingtième siècle" *Revue d'histoire moderne et contemporaine*, 14 (October-December 1967), pp. 375-405. For a history of peasant unrest since colonial times, see: *Allpanchis*, Vol. XI (Cuzco, 1975), *Movimientos campesinos*.

15. These figures are very similar to those collected in the early 1960s by: Comité Interamericano de Desarrollo Agrícola (CIDA), *Tenencia de la tierra y desarrollo socio-económico del sector agrícola: Perú* (Washington: Pan American Union, 1966), p. 56 and 107. However, the earlier CIDA study greatly understated the importance of communally held pasture lands.

16. Fernando Eguren,"La Tierra: Su Distribución y Los Regimenes de Tenencia" (Lima: Unpublished paper, Desco, 1978). In the officially registered highland peasant communities, technically all land was communally owned and was, therefore, registered as a single unit. Over 30 percent of all sierra agricultural land belonged to some 1,200 registered communities with a portion of that land farmed as individual subsistence plots and most used as communal pasture. Communal land was not a significant factor on the coast.

17. Sometimes this led to arrests, violent expulsions from the land, or confrontations with the police. At other times the peasants won, usually after having formed a *sindicato* (union). See: Cotler, "Traditional Haciendas..."; LaMond Tullis, *Lord and Peasant in Peru* (Cambridge, Mass.: Harvard University Press, 1970).

18. García-Sayán, op. cit.; Eguren, op. cit. Of course, even with the land owned by peasant communities factored out of Table 1, nearly half the land in the sierra (about 9 million hectares) and 60 percent of the coastal agricultural area were still under the control of large haciendas (over 200 hectares). But Eguren and García-Sayán's analyses show that increasing numbers of sierra peasants were changing from neofeudal peons to cash-paying tenants or owners. Hired labor was also becoming more common in the Andes, another aspect of the capitalist transformation already taking place.

19. The 1960s mobilization of Andean peasants is more extensively discussed in: Eric Hobshawm, "Occupaciones Campesinas de Tierras," *Analisis* 2-3 (Lima, April 1977), pp. 111-147; Handelman, *Struggle* Chapters 5-6; Hugo Neira, op. cit.; Alberto Flores Galindo, "Apuntos sobre las ocupaciones de Tierros y el sindicolismo agraria, 1945-1964" *Allpanchis* XI (Cuzco, 1978).

20. For excellent analysis of the agrarian reform: José María Caballero, "Reforma y Reestructuración Agraria en el Perú" (Lima: CISEPA, Universidad Católica, 1976); Mariano Valderrama, *7 Años de Reforma Agraria Peruana* (Lima: Universidad Católica, 1976); Douglas Horton, *Land Reform and Reform Enterprises in Peru* (Madison, Wis.: University of Wisconsin Land Tenure Center, 1974).

21. *El Comercio* (Lima, October 12, 1969).

22. North, "Political Conjunctures..."

23. John Gitlitz, "Impressions of the Peruvian Agrarian Reform," *Journal of Inter-American Studies and World Affairs*, 13 (July 1971), pp. 456-58.

24. Valderrama, op. cit., p. 51, and Caballero, op. cit., p. 12. The amount of land which hacendados could maintain exempt from the agrarian reform was an extremely controversial issue and was periodically lowered or raised as the result of competing peasant and landlord pressures.

25. See: Eric Hobsbawm, "Generals as Revolutionaries," *New Society* (November 20, 1969); Gitlitz, op. cit.; Valderrama, op. cit.

26. Of course the military did conceptualize the agrarian reform partly in political terms in that they hoped to co-opt future rural unrest and thereby depoliticize the countryside. In fact, the reform had the opposite effect.

27. *El Comercio* (April 27, 1971), p. 4.

28. Hacendados were paid in long-term bonds, with land value often determined on the basis of the owners' (vastly undervalued) earlier tax statements. On the financing mechanism, see: J.M. Caballero, "Aspecto Financieros en las Reformas Agrarias" (Lima: CISEPA, Universidad Católica, 1975).

29. These figures exclude land in the sparsely populated jungle region. Data on the amount of land under agricultural or pasture utilization in Peru vary widely. The generally accepted figure is 19-20 million hectares for the coast and highlands and perhaps 23 million hectares including the jungle. But the Agrarian Reform Agency has issued figures as high as 29 million. See: Caballero, "Reform y Reestructuracion...," p. 7.

30. Excellent extended accounts of hacendado and peasant mobilization (from 1970-1976) related to the agrarian reform are to be found in: Valderrama, op. cit.; Henry Pease García et al., *Estado y Política Agraria* (Lima: Desco, 1977); Colin Harding, "Land Reform and Social Conflict in Peru," in Lowenthal, ed., op. cit.

31. *La Prensa* (June 25, 1970).

32. The percentages of delegates designated by the government varied according to the CAP's debt to the Agrarian Reform Agency. See *Valderrama*, op. cit., p. 56.

33. *La Prensa* (April 27, 1972).

34. Quoted in Abdón Palomino, "Andahuaylas, 1974: Un movimiento de reivindicación campesina dentro del proceso de reforma agraria," *Allpanchis* (Cuzco, November 1978).

35. Adjudicated land is property in which title has been transferred to peasant enterprises (CAPs, SAIS) or individual peasants. The gap between area expropriated and area adjudicated (i.e., those lands still held by the Agrarian Reform Agency) narrowed considerably from 1976 through 1978. Note that the final adjudicated area is expected to exceed the area expropriated because 250,000 hectares of public land will be included.

36. The 21 million hectare figure reflects expansion (from 1964-1974) in the number of hectares under exploitation.

37. Again, these data reflect titles actually transferred into peasant hands. The figures include some land adjudicated from 1967-1969 under the 1964 agrarian reform law, but this quantity was minimal. Social Property is another form of cooperatively controlled enterprise. It was supposed to become a central feature of rural and industrial production as envisioned by the more radical Velasco government but was quietly phased out under President Morales Bermúdez's move to the right.

38. Caballero, "Reforma y Reestructuración..."

39. Some differences do remain. The CCP's leadership is more ideologically homogeneous (almost exclusively Marxist) and has more extensive links to urban leftists. The CCP's rank and file is self-selected and clearly represents atypically militant peasants. The CNA, on the other hand, has many local leaders who are fairly moderate and progovernment. The rank and file is more diverse and far less uniformly militant.

40. The dissolution of the CNA's national-level organization followed close on the heels of the government's March 1978 announcement that SINAMOS would be abolished by August of that year.

41. Like CCP officer Hugo Blanco, Avelino Mar was elected to the national Constituent Assembly in June 1978 (only one month after he was charged with conspiracy) and now enjoys congressional immunity.

42. Conversation with Mr. David Gow. For a discussion of racial tensions and the prejudices of rural bureaucrats in the Peruvian sierra, see: Charles Kleymeyer, *Social Interactions Between Quechua Campesinos and Criollos* (Madison, Wis.: University of Wisconsin Ph.D. Dissertation, 1973); David Gow, *The Gods and Social Change in the High Andes* (Madison, Wis.: University of Wisconsin Ph.D. Dissertation, 1976).

43. Harding, op. cit., p. 220; Horton, op. cit., pp. 78-79.

44. Caballero, *Reforma y...;* Data were drawn from Douglas Horton, Peru: *Case Studies Volume* (Washington: IBRF Studies in Employment and Rural Development, #22, 1975).

45. Eguren, op. cit.

46. As we have seen, there were multiple motivations for the agrarian reform, not all of which were related to peasant pressures. However, the desire to pacify the countryside was an important factor.

47. Quoted in North, "Political Conjunctures...," p. 27.

48 . A recent bright spot in the Peruvian export picture has been the sharp rise in coffee prices. Copper prices too have recovered somewhat from their post-1974 decline. But Peru's export revenues from metals have been badly hurt by crippling miners' strikes. The fishmeal catch has also rebounded somewhat after a sharp decline and the price of sugar has been up in 1979-80. In all, export revenues are still inadequate when compared to the nation's indebtedness.

49. United States Agricultural Attaché, *Situation Reports* for 1976-1978.

50. Eguren notes that 87.5 percent of latifundia land was devoted to pasture, while 87 percent of Peru's entire crop land was held in units of under 50 hectares (and, thus, not subject to agrarian reform). See, Eguren, op. cit.; García Sayán, op. cit.

51. Webb, op. cit.; Richard Webb and Adolfo Figueroa, *Distribución del Ingreso en el Perú* (Lima: Instituto de Estudios Peruanos, 1975).

Agrarian Reform and the International Consensus

Alan W. Horton

Impelled by self-interest and a sense of mission, industrialized countries and various international agencies have for several decades been giving financial aid to developing countries for projects and programs that donors and recipients alike have hoped would facilitate the process of development. Accordingly, at any given time, there has been a prevailing theory or consensus on what development consisted of—its goals, its methods, how best to spend money to induce it. Only in the past half decade has the consensus shown signs that rural development ought to play a more central role in effective overall development.[1]

But if the central importance of rural development is increasingly a part of the international dialogue, clear differences of opinion on what constitutes the most effective kind of rural development are apparent. What kind leads to greater national affluence and well-being? What kind promotes greater equity? A central issue in this debate has revolved around the notion of agrarian reform.

The phrase "agrarian reform"[2] has had various meanings, ranging from the simple redistribution of land to sophisticated provisions for the political, social, and psychological impacts of any radical restructuring of rural life. Raup says that "land reform is...a basic change in land tenure arrangements—in relations among men with respect to land—together with supporting measures necessary to achieve its objectives."[3] In general, despite attempts by some governments to water down its meaning and use it as a palliative, the phrase has retained its revolutionary implications—a drastic reordering of land tenure systems in the search for greater equity. The Mexican and Russian Revolutions defined it this way. So, inferentially, did the American forces of occupation after World War II in Japan and Taiwan. Less drastic was the definition given by the Egyptian Revolution of 1952; dramatic actions in the name of agrarian reform broke the power of the absentee landlord class but finally did little to change the actual relationship of peasants and land.

Despite its position in major revolutions of this century and its pivotal importance in various countries after World War II, agrarian reform has had a somewhat happenstantial relationship to orthodox development theory over the past several decades.[4] On the other hand, the relationship of rural development to development theory has been constant if sectoral, peripheral to development's main thrust but supportive in its sectoral way. Off in their special corner of the intellectual universe, both agrarian reform and rural development have been topics for discussion, refinement of thought, and missionary zeal among sectoral bureaucrats—such as the staffs of the United States Department of Agriculture (USDA) and the Food and Agriculture Organization of the United Nations (FAO), academics, and others. But when they have produced ideas for the consideration of the mainline theorists of development, the ideas have usually been absorbed into a general concern to achieve national production targets and boost per capita growth rates.

In the post-World War II era, among those who argued about the role of agriculture, two areas of belief, often overlapping, are discernible. The first, as already indicated, has to do with the conviction that rural development should be central to an effective overall theory of development; FAO, for example, has long sought to escape from its sectoral cage and play a greater part in formulating the strategies for development decades. The second area of belief relates to the conviction that rural development, sectoral or central, will not be successful without agrarian reform. The ideological component in the second belief is strong, and from time to time agrarian reform has thrust itself into public view, breaking out of the entanglements of rural development theory and standing on its own against the skyline of revolution.

* * * * *

It is instructive to look at some of the interplay between notions of development, agrarian reform, and other concepts during the three and a half decades since World War II. In 1951, the Madison campus of the University of Wisconsin sponsored and hosted the International Conference on Land Tenure and Related Problems in World Agriculture.[5] Funded largely by the Technical Cooperation Administration and the Mutual Security Administration of the United States, it lasted 7 weeks and brought together 3 representatives from each of 40 countries (excluding the Soviet bloc countries) to report on their land tenure problems and discuss the need for comprehensive programs of agrarian reform in the world's various regions. While it did not ignore other aspects of rural development, such as the need for new roads and drainage systems and better agricultural training and research, nonetheless its major preoccupation was the developmental wisdom and distributional justice of land reform.

The final impact was expectably small. Sweeping change is seldom the result of international conferences, however fervent the call to arms. Indeed, after the reforms induced by war and other upheavals in the late 1940s in Eastern Europe and East and South Asia, there was a land reform lull. Agrarian reform furled its revolutionary banners and—despite events in Egypt (1952) and Bolivia (1953)—resumed its place in the literature as one of the obedient handmaids of rural development. After the Wisconsin conference, a number of persons who had attended went almost directly to a meeting of the FAO governing body and were among those who arranged the passage of a resolution to the effect that FAO should take upon itself the study of agrarian reform. The resolution expanded the definition of agrarian reform to include such problems as marketing, agricultural credit, rural taxes, and the purchase of supplies. In one sense, this expansion of the definition can be seen as a logical response to the growing complications caused by new agricultural technology.[6] In another sense, it can be seen as an effort by the forces of moderation, in the name of rural development, to encompass and control the ideological and radical thrust of agrarian reform.

Development theory of the early and middle '50s focused largely on technology and "know-how," and later in the decade, when the evidence suggested that technology was not enough, the notion of investment (the more massive, the more effective in production terms) was added. Rural development fit easily into these perceptions—as a sectoral supplement. The sector should, according to theory, do its part to add to the totality of national production, with an emphasis on agricultural production for export. Agrarian reform was increasingly seen as merely one of several incentives that might be used to improve the agricultural sector's contribution to the whole; its call for radical rural change became only one voice in a chorus of calls for technologically "sound," production-oriented measures in the countryside.

In 1960, when the proposals for action in the first development decade were drawn up, new notions were introduced. Production and investment, it was noted, are important to the development process but "there is now greater insight into the importance of the human factor in development, and the urgent need to mobilize human resources."[7] Stress was laid on good national planning as the way for each nation to advance, and the approaches to planning should be comprehensive, including social and cultural as well as economic factors. Community development in the rural areas was cited as one example of the comprehensive approach and as an activity of coming importance; it was described as "a method and an approach which is applicable to the problems of rural development as a whole, including the key element of land reform. Land reform, in this context, is much more than changes in land tenure or redistribution of land ownership; it is in the widest sense of the term synonymous with agrarian institutional reform."[8]

The 1960 proposals for action also foreshadowed other notions and practices of later years. For example, tucked away in the document in one place or another are most of the then-existing ideas on development, as if each UN agency or each pressure group had worked its own notions into the document at whatever cost to the internal consistency of the whole. Radical bits of theory, such as agrarian reform, are put into intellectual isolation in the middle of otherwise conservative paragraphs. Yet the resulting theoretical potpourri exudes a spirit of universal consensus that hovers over and helps hide its discrepancies—a fact that was doubtless pleasing to those who believe that ambiguity can be politically productive. In later years, as rhetoric improved and internal inconsistencies became fewer in international documents, the consensus-seeking habit remained, and it has influenced the ways in which developed and developing countries behave toward each other, as I hope to show.

Though the document is internally inconsistent, its treatment of rural development and agrarian reform is strikingly consistent with a great deal of past practice. The two concepts are kept in their sectoral place. Appearing, ironically, in the agricultural section are several pleas that agriculture should play a more important role in the planning process. Aside from a discussion of rural areas under the rubric of community development, the also-there status of the peasant is perpetuated. Significantly, in his five-page foreword, Acting Secretary-General U Thant did not mention the word "agriculture" or use any phrase that referred directly to rural matters.

But in 1961, in the rhetoric of the Alliance for Progress, rural development did achieve a more central place—led there, significantly, by a surge of interest in Latin American land reform. In the "Declaration to the Peoples of America" that precedes the Charter of Punta del Este, the following paragraph is given prominence.

To encourage, in accordance with the characteristics of each country, programs of comprehensive agrarian reform, leading to the effective transformation, where required, of unjust structures and systems of land tenure and use; with a view to replacing latifundia and dwarf holdings by an equitable system of property so that, supplemented by timely and adequate credit, technical assistance and improved marketing arrangements, the land will become for the man who works it the basis of his economic stability, the foundation of his increasing welfare, and the guarantee of his freedom and dignity.[9]

Other parts of the declaration and the Charter contain complementary specifications for what amounts to a blueprint for revolutionary change. The Charter was signed by the United States and all Latin American countries except Cuba.

The impact of the Alliance for Progress is a matter of dispute. An American diplomat who was involved says that it established the climate for most of the reformist changes that have occurred since 1961; a Brazilian civil servant also interviewed by this author says its principal contribution was to frighten Latin America's conservative forces to the point that genuine reform became virtually impossible. In the 1950s, according to one writer, Latin America had "emerged as an area in which land redistribution was most needed—and most vigorously debated."[10] The Cuban Revolution of 1959 added a note of urgency for those who wanted to use reform to avert further revolution in the hemisphere. Another writer adds that the Alliance for Progress made the debate a respectable one and "discredited those politicians who argued that all programs of land redistribution were Communist-inspired... and it probably spurred several countries which otherwise might have stood still to undertake at least a beginning."[11] The United States contributions were in the form of program aid, as opposed to project aid, and were awarded on the basis of country planning that appeared to meet the right specifications.[12] Despite the contributions, however, land reform programs in the Latin America of the 1960s were quiescent, even inert, in all but a handful of countries.[13]

Of particular interest is the commitment to land reform and rural revolution that Latin American countries made at Punta del Este. Many of the regimes that signed the Charter clearly did not intend, at least in the light of subsequent events, to put the rhetoric to the test of reality. Presumably, at a time when notions of reform were becoming stronger throughout Latin America, it was politically expedient to be associated with the Alliance, and presumably the possibility of massive United States aid was a powerful attraction. Whatever the reasons for signing, Latin American governments became part of a consensus, or rather consensus-in-rhetoric, joining a major donor government in a statement of developmental intent. Comparing the Charter to the UN proposals for action for the first development decade, the principle of establishing a consensus-in-rhetoric remains the same but the rhetoric of the Alliance is more internally consistent and more radical. And the gap between rhetoric and reality is wider.

The gap widened, of course, as idealism increased. While notions of equity had always been sprinkled about in the literature and had even been dominant on special occasions such as the Wisconsin conference of 1951, in the 1960s they became a major theme of, and even an organizing device for, a great deal of development rhetoric. The goal of development, by implication, became fairness to the have-nots within developing countries as much as it was national affluence. As at Punta del Este, agrarian reform was the natural spearhead for the effort to be fair to peasants.

A growing concern about rural equity led to the World Land Reform Conference of 1966. Sponsored by both the UN and the FAO, it brought together some 225 persons nominated by 76 governments plus a group of about 100 specially invited participants, officials of the UN system, and representatives of private organizations. The stated purpose was "to provide an opportunity for the exchange of experience on a worldwide scale...in order to assist Governments in their assessments of agrarian policies as related to the planning and implementation of structural reform programs."[14] During the two weeks of the conference, three working parties wrestled with the following topics: (1) problems of land tenure and structural reforms; (2) social and economic aspects of land reform; and (3) administrative, financial, and training aspects of land reform. An entire range of points of view on each topic is reflected in the conference report, which nonetheless shows a praiseworthy internal consistency. Verbal escape hatches, expectedly, allow for the special circumstances faced by individual governments, and hence nothing that might be construed as commitment stood in the way of the unanimous adoption of the working-party reports and the final resolutions. The report lays special stress on the productive aspects of land reform and its special place in successful development programs. In one of the opening addresses, Gunnar Myrdal dampened some of the ideological fires by reminding conferees of the unproductive results of indecisive action: "It is the worst of both worlds to have a radical ideology of 'land to the tiller' and a largely ineffective legislation conforming to that ideology...."[15]

Despite the emphasis on the potential of greater productivity if agrarian reforms were adopted, it is the rhetoric of equity that sets the tone of the 1966 conference report. Significantly, some of the rhetoric explored the relations of developed and developing countries. In an analytical mood, Myrdal anticipated that "the rich countries will find it necessary to press for reforms in the underdeveloped countries; and the latter countries will be forced to press for aid from the rich countries to enable them to carry out the reforms.... More reform in the [under developed] countries will make aid seem more meaningful in the rich countries, while more aid will increase the scope of possible reform."[16] Less analytically, but presaging things to come, Hernan Santa Cruz, the chairman of the conference, spoke in his closing address of the "insistent demands for greater equality and social justice. The disparities of income and wealth between the industrialized and the underdeveloped countries are increasing day by day. This pattern carries over to underdeveloped countries, where there is a tremendous gap between conspicuously wealthy minorities and the poverty-stricken masses."[17] This coupling of injustice between nations and within nations has now become commonplace on the agendas of international forums.

It is difficult to assess the impact of the 1966 conference, or any such conference. Like the Wisconsin conference of 1951, it was a specialized gathering, reflecting the sectoral concerns of those immersed in or well-acquainted with agricultural affairs. A principal theme, as before, was the crucial role of agrarian reform for both rural and general development. Though the conference may have created a momentary flood of interest in rural affairs, it did not succeed in finding for rural development a central place in development theory. In terms of consensus-in-rhetoric and its relation to reality, the gap was as wide as the ideological component was strong.

At the launching of the Second Development Decade in 1970, the rhetoric had changed somewhat but not its place in rural development. The General Assembly resolution makes only one direct assertion about agrarian reform and rural development. Under the section entitled "Expansion and diversification of production," the following appears:

Developing countries will formulate, early in the Decade, appropriate strategies for agriculture... designed to secure a more adequate food supply..., to meet their nutritional and industrial requirements, to expand rural employment and to increase export earnings. They will undertake, as appropriate, reform of land tenure systems for promoting both social justice and farm efficiency.[18]

Clearly, the agriculturalists at the 1966 conference had not succeeded in persuading others that rural affairs were more than sectorally important.

The notion of equity still pervades the General Assembly resolution of 1970, but the emphasis is different from that of 1961. It is less on the poor of developing countries than it is on the unfair imbalance between rich countries and poor countries. While calling for development and asserting that the responsibility for it is in the hands of developing countries themselves, the preamble reflects a preoccupation with "international cooperation for development"—by which it plainly means more development aid, more effective transfer of technology, and above all better terms of trade for the developing countries. This emphasis had been emerging in other international forums, notably meetings of the United Nations Conference on Trade and Development (UNCTAD), and it had become a central poor-country concern. Having achieved voting majorities and having formed a voting bloc (the Group of 77) at international meetings, the developing countries now began to define their interest in international dialogue in terms of that concern. This approach to dialogue continues today in all international forums, whatever the formal topic.[19]

Thus, before the so-called food crisis of 1973-74, both developed and developing countries had, with varying consistency, adopted postures and strategies that required of the other group a radical change of attitude, if not action. Both were sporadically insisting on their own definitions of justice.

* * * * *

The World Food Conference of 1974 was held in an atmosphere of crisis. When the conference was first suggested in 1973, the harvests had been excellent; when it finally met in Rome in late 1974, bad harvests had created a food disaster.[20] Developing countries were under pressure to consider publicly their internal problems related to food production. If agrarian reform promoted greater food production, why should developing countries not act accordingly? And developed countries, not only the food-exporters but all who might provide external aid for local food production or participate in trade schemes to increase income available for agricultural investment, were faced with the necessity of exhibiting a sense of global responsibility and of responding as if the world were as interdependent as all had been saying. The atmosphere of crisis was deepened by the success of OPEC in quadrupling the price of oil and the expectation that the price would rise further. For both developed and developing countries, the prospects were dismaying.

At the same time, the insistence of the developing world on new terms of trade was crystallizing. Trade patterns for food and other commodities had already been the subject of considerable international discussion, most clamorously at UNCTAD meetings. In early 1974, the UNCTAD dialogue and other factors (such as the heady atmosphere brought on by the success of OPEC) led to the adoption—with careful reservations by most developed countries and outright disagreement by six of them—of two General Assembly resolutions that are now referred to as the New International Economic Order (NIEO). The NIEO has set a standard for the demands of developing countries and is a blueprint for what they think developed nations should agree to in order to improve the trade position of those less fortunate.

The agenda of the World Food Conference reflected the concerns of both groups, with the NIEO getting its share of discussion. Sponsored not by the FAO but by the United Nations itself, the conference based its deliberations on the notion that the issues to be confronted were political and economic rather than technical and that "firmer political wills were needed to commit the increased resources and make the policy decisions essential for a more rapid increase in food production."[21] It was attended by ministers from all parts of the globe, including many who were *not* ministers of agriculture; unlike other international conferences on food affairs, it received extensive coverage in the world press. The debate dealt with matters of food production, food security, and food trade. The final resolutions represented rhetorical trade-offs between developed and developing countries; there were alternating exhortations for concessions on food trade and aid and for internal reforms as incentives to greater food production. One example of such alternation will suffice: food trade should be expanded, liberalized, and stabilized, but a fair share of the increased national income of developing countries should go to rural populations.

The results were mixed. In terms of food trade, agreement was minimal and the whole matter was plainly in the realm of continuing dispute. In terms of food production, however, the conference resulted in a marked increase of external aid to food-deficit countries.[22] In terms of food security, good beginnings were made for global planning, including the creation of a negotiating arena for the planning process in the form of a Committee on World Food Security lodged within the FAO. Moreover, several other new international institutions were created, two of which have survived. The first is the World Food Council (WFC), which is really an arm of the UN Secretariat, and is striving to establish itself as a security council for world food affairs, a ministerial-level mechanism for coordinating political decisions about food. The second is the International Fund for Agricultural Development (IFAD), a separately and amply funded specialized agency in its own right, which provides development aid for agriculture, particularly to those countries and to those populations that the President of the World Bank had spoken of in the early 1970s, namely, the "poorest of the poor."[23] This proliferation of international "food organizations" reflects a new preoccupation, which has also brought about a proportionally greater investment in food production on the part of such older organizations as the World Bank and the United Nations Development Program (UNDP). The presence of the two new organizations in Rome, where they are to some degree hosted by FAO, has created an atmosphere of careful rivalry in both rhetoric and action. Some observers interpret the existence of WFC and IFAD as a reprimand to FAO, a signal from the World Food Conference that FAO had been sticking unimaginatively to its technical jobs without meeting the challenges of the times.[24]

With regard to the international consensus, whether for rhetoric or action, the significance of the World Food Conference is its invitation to agriculturalists to come out of their sector and take part in development dialogue as full partners. The agenda of any discussion of general development at an international meeting now devotes some prime time to problems of rural development. Understandably, ambivalent feelings remain about the role of agrarian reform; political structures are conservative. But insofar as ideas about it are entangled in theories of rural development, agrarian reform is now bound to receive more debating time—even without the revolutionary thrust that gave it such prominence at Punta del Este.

Indeed, food matters have become so preoccupying to governments that they have become an important instrument of international relations. Since the World Food Conference, governments have been unable to avoid knowing that a real food problem exists, and they have become uncomfortably aware that control of the problem is becoming more difficult year by year. The projections, even at their most optimistic, are deeply discouraging.[25] In this situation, developing countries, giving the appearance of deft maneuver, have come to recognize and use the leverage value of food. In the words of a Third World delegate to numerous international conferences, "food is now the weak man's instrument." Under the banner of food for the hungry, developing countries can use strong, indignant language without fear of appearing to whine, and they can put donor countries on the defensive. Food, it turns out, is an effective shield in the battle to move toward the NIEO; what better way to get at problems of international trade than through publicity-sensitive debate on the trade of foodstuffs and other raw materials between rich countries and poor countries?

* * * * *

Since 1974, there have been many international meetings of one sort or another in the UN system. Some have been for specialists, some have been high-level. But all have used the issue of equity to score rhetorical points and move the climate of opinion in the direction of (1) more internal social justice, especially in rural areas, or (2) better terms of trade and more aid. The assumption is that the international climate of opinion or consensus, or the intersection of two climates of opinion, will determine in some measure the definition of development under which cash transfers occur.

A case in point is the World Conference on Agrarian Reform and Rural Development (WCARRD), which took place in July 1979. It was sponsored exclusively by FAO and was, in fact, seen by some as an FAO attempt to regain the initiative it lost at the World Food Conference. WCARRD was a full-dress affair with the customary over-ample documentation, long statements of country positions—chiefly for home consumption—at a plenary session that lasted the entire two weeks, and representation from almost every country in the world. Many delegations were large, often led by ministers of agriculture; in some instances delegations were led by distinguished public figures, such as Andrew Young of the United States. Several heads of state and the Pope came to address the conference.

The result was a "Declaration of Principles and Programme of Action,"[26] which had been initially drafted by the FAO secretariat and revised slightly and approved first at the level of a preparatory meeting several months earlier and then at the plenary. The detailed work of getting the approval was done at the conference by two commissions. In conformance with the accepted pattern for debate, one commission dealt with the section entitled "National programs of action in developing countries," and the second dealt with "International policies for agrarian reform and rural development." The importance of this pattern was candidly acknowledged in private by an FAO civil servant before the conference took place; the developed countries, he said, were interested in coming because they wanted to discuss internal reforms in the developing countries, and the latter wanted to discuss the NIEO in the guise of discussing the trade of food commodities. The compromise, he hoped, would be that each side would listen to what the other wished to talk about and give a little on previous positions. In the meantime, he asserted, the job of the FAO as the middleman in the great debate was to stress and publicize the central role of rural development in the planning process. He might have added, but did not, that if the publicizing was successful, the lead role of FAO in rural affairs would be reaffirmed at a time when other members of the UN family were clamoring for greater portions of the new action made possible by the new interest in food.

As expected, in the commission on international trade the developed countries showed their willingness to discuss the terms of trade and aid and continue the NIEO debate from other forums. As usual, they entered numerous reservations at the relevant parts of the text of the program of action, watching carefully lest they might be called upon at some future date to translate words and postures into funding commitments. An American delegate compared the concern with wording of such documents to the common law of the Anglo-Saxon tradition. If, he said, his delegation did not pay careful attention and, say, out of sheer weariness permitted without specific dissent the inclusion of unacceptable phrases, the phrases would then be repeated in subsequent documents and could become part of "a generally accepted body of international opinion." Correspondingly, of course, the expectation of the Group of 77 is that each international forum, under continuing pressure of attack from behind the shield of food and other inequities, will produce minor changes in conventional wording that will finally add up to something financially useful.

The developing countries took a different defensive tack. With reservations from only a handful of countries, they agreed to a text on national policies that implied approval of revolutionary changes in internal power structure. In interviews with more than a dozen Third World delegations, it was plain that agreement on the text did not, in their view, commit their governments to any particular line of action. Consider the following plea:

> *In countries where substantial reorganization of land tenure and land redistribution to landless peasants and smallholders is needed as part of the rural development strategy and as a means to redistribution of power, governments* should consider action to *implement redistribution with speed and determination backed by preventive legal sanctions to avert disinvestment and evasive transfers*... (emphasis added).[27]

They were being asked, they said, to consider actions; to say they would consider them was a small price to pay for the chance to talk about international trade. Among the delegations of the developed countries was a tired acceptance of this gap between the home performance of developing countries and the rhetoric they gladly espoused, but it was also possible to detect among those delegations a faint hope that one day, perhaps this time, rhetoric would have an effect on action. Plainly, however, delegates who agree to international documents like that of WCARRD cannot be called to account. The crunch does not come until the moment when specific projects are negotiated between donor and recipient, the moment when the country with the money can influence action.

To what extent can the country with the money influence action? Theorists of development disagree.[28] After Punta del Este, when the United States made money available on the basis of approved and detailed national plans, hopes were high, but the result tended to be rural development rather than the agrarian reform called for in the Alliance charter. This experience, it appears, is not unusual. Idealistic rhetoric beforehand notwithstanding, donor countries and donor agencies find it hard to give money for projects that help the "poorest of the poor."[29] Some would claim the donors do not want to do so—because of their regard for the power structure in the recipient country, because of power structures among donors, or because in a world of nation-states it is the course of least resistance. One seasoned negotiator from the Third World said that if the World Bank took McNamara's earlier pronouncements on equity too seriously, no World Bank projects would exist.[30] On the other hand, even fitting in with a national plan that seemingly favors the continuance in power of a dominant elite, a project or series of projects can provide a crucial push in reformist directions or can even have unforeseen spin-offs that accelerate greatly the process of social and political change. Delegates at WCARRD expressed a wide range of opinions on the matter, but most donors and recipients agreed, recognizing the conservative implications of what they were agreeing to, that the right of governments to make their own final selection of projects available should not be infringed by the power of other nations.

It would be wrong, of course, to give the impression of a seamless uniformity among either developed or developing countries. Among nations there is an intricate set of networks that does not depend on membership in the OECD or the Group of 77. If there is a special form of dialogue for North-South relations, the same is true for the confusion of other relationships that exists between and among national and cross-national entities. These other relationships create significant divisions and linkages within and between each camp. Equally significant, however, is the growing determination on the part of developing nations to adopt a common strategy at international forums—despite other pulls and pushes. At WCARRD, the Group of 77 held special meetings, and the Group's current chairman, recognized as such by the conference chair, often spoke on behalf of other Third World nations.

If the adoption of a common strategy toward the other world should not obscure the divergence of beliefs within each world, the same holds true for beliefs within each country. A wide range of opinion about agrarian reform, for example, exists in the United States. In the view of some, it interrupts the building of an infrastructure on which rural development must finally depend, or it prevents the emergence of commercial agriculture on whose efficient methods the world must finally depend; in the view of others, it promotes the production of food and fosters food independence, or it is crucial for the establishment of an equitable society. The same range of belief, though not always articulated perhaps, exists in many developing countries. At WCARRD, some members of some delegations from countries with highly inequitable land systems expressed themselves as sincere believers in agrarian reform and in off-the-record interviews allied themselves with the forces of change at home. Some stated explicitly that the elites of the developing world, presumably including themselves, are the principal obstacles to development; echoing recent literature of dissent in donor countries,[31] they suggested that the effectiveness of development aid has not only regressed in recent years but that development aid is in fact counterproductive because it sustains elites in power and prevents the revolution or rapid evolution that would bring about a fairer distribution of national wealth. At one point the British delegate to one of the commissions, referring obliquely to such notions, ruffled many developing country delegates by stating that his government wanted "value for money" and had it in mind to provide less aid on a government-to-government basis and more aid by way of voluntary agencies; several Third World representatives of my acquaintance gave their silent approval to this threat to by-pass elitist influence.

Indeed, though they were all aware of the gigantic difficulties involved in planning radical social change, several developing country delegates were outspokenly self-critical. The high point of rhetoric, if not awareness of realities, was a speech on the second day by Julius Nyerere of Tanzania. He divided his speech into the two customary categories and spoke with great candor of the disparity of rich and poor between developed and developing nations and within developing nations. About the latter inequity, asserting that the poor of the developing world are largely rural, he said:

Governments by themselves cannot achieve rural development. They can only facilitate it and make it possible. They can organize, help and guide; they cannot do. For rural development is people's development of themselves, their lives and environment. And the people cannot do it if they have no power....[32]

And then, quoting Michael Lipton (*Why Poor People Stay Poor*) against himself:

The sincere egalitarian rhetoric of, say, Mrs. Gandhi or Julius Nyerere was, allowing for differences of style and ideology, closely paralleled during early industrial development.... But the rural masses of India and Tanzania lack the power to organize the pressure that alone can turn such rhetoric into distributive action against the pressure of the elite.

Later in his own words:

If the people are to be able to develop they must have power. . . . At present the best intentioned governments—my own included—too easily move from a conviction of the need for rural development into acting as if the people had no ideas of their own. This is quite wrong. At every stage of development people do know what their basic needs are. And just as they will produce their own food if they have land, so if they have sufficient freedom they can be relied upon to determine their own priorities of development and then to work for them.

Still later:

And finally, but underpinning everything else, effective decision-making power—political and economic power—must lie with the inhabitants of the rural areas. Political power has to be exercised by the poor if the present flow of wealth towards wealth is to be stopped. . . .

Elsewhere in the speech, President Nyerere drew a parallel between poor rural areas and rich urban ones within countries, on the one hand, and poor developing nations and rich developed ones, on the other. He implied, though did not say specifically, that if rural areas need political clout to protect themselves and make certain of their rights, so do developing nations. One Third World delegate commented later: "The trouble is that the only clout we have comes from OPEC and a voting majority in UN institutions."

The matter of political clout carries the equity arguments one step further. Political clout emerges as one of the ways to achieve equity and perhaps the only way to maintain it. To give genuine political power to peasants is, in most developing nations, more of a revolution than agrarian reform—it is agrarian reform plus a preservative. In Egypt, for example, it is unthinkable that real rural power should pass from the hands of government officials, despite the rhetoric of land reform that has been a part of Egyptian life since 1952; urban elites will not be giving up their prerogatives in the countryside to the peasantry. Nor will rich countries be giving up more than they must to poor countries. It will be almost as hard to sell the NIEO to the electorates of the developed world as it will be to persuade urban elites of the developing world that peasants can manage their own affairs in the national interest. The problem is that each debating team is asking for something that goes against the grain of the other's system and style—and the strains will grow as the rhetoric improves.

The overlap of the two positions defines the liberal stance in both instances. Developed countries, say the liberals of the developed world, should make concessions on international trade, agreeing to what amounts to international legislation for equity. Developing countries should, at the least, arrange for projects that reach "the poorest of the poor." Most bureaucrats of the UN system find themselves in this overlapping area as they seek to act as international brokers. Indeed, it is this overlapping area in which the international consensus is lodged and institutionalized by the UN system. As compromises and trade-offs are sought, the consensus sets guidelines. It provides directions of change for groups of activists and revolutionaries. It sets standards for action leading to greater equity, for what each country ought to be doing about its leaders. As developed and developing countries seek the approval of the community of nations—and in some measure all of them, or factions with them, do—the consensus by its very existence exercises what Soedjatmoko of Indonesia has referred to as a "sanction function." When developing nations seek development aid, they use the verbiage of the current consensus to justify the request—and donor countries or international agencies, whatever their political motives, respond in kind.

Yet few of the delegates at WCARRD saw anything good coming out of the conference. While some were pleased that the need for rural development was receiving effective publicity, most were persuaded that the rhetoric would not lead to significant action. They discerned not a consensus but an obstinacy on the part of the delegates of the other camp. But the consensus exists in the minds of some, and it has changed since 1974. Rural development is "in" as a major factor in general development, and agrarian reform has become more respectable. Its increasing appearance in consensus circles will in some measure affect political climates in the Third World. A proposition at WCARRD may also affect that climate; FAO, it was resolved, should take the lead in monitoring Third World progress in rural development and agrarian reform. Such monitoring would entail public judgments, and if political circumstances were to permit a modicum of honesty beneath the wordy diplomatic overlay that FAO publications occasionally exhibit, liberal positions in the developing world would be strengthened.

As this is written, the rhetoric for the resolution on the Third Development Decade is being negotiated. If consensus-in-rhetoric can sometimes affect the climate of decision about internal reform, trade, and aid, the resolutions should provide clues for the near future. Will the role of rural development be stressed, and what recognition will be given to the role of agrarian reform? First indications are that greater food production will receive major attention. The next question is: Can more food be produced and greater equity be achieved simultaneously? The consensus says this ought to be possible; the sticking point is the slow pace of reality.

(September 1980)

NOTES

1. I am particularly grateful for comments on an earlier version of this paper by Barbara Huddleston of the International Food Policy Research Institute and by Howard Handelman, Director of the AUFS Politics of Food Project and Associate Professor of Political Science at the University of Wisconsin in Milwaukee.

2. Or its near-twin "land reform." Though some make distinctions between the two phrases, I am not making them here. Agrarian reform is sometimes seen as the more inclusive phrase, including not only land reform but the other measures necessary to make it effective socially and economically.

"Rural development," on the other hand, is used here as a catchall category that subsumes not only agrarian reform but any other measures (radical and nonradical) that make rural areas more productive. Rural development (in the form of the building of roads and dams or the improving of extension services, for example) can occur without agrarian reform.

3. Philip M. Raup, "Land Tenure," *Encyclopedia of the Social Sciences* (1968), p. 571.

4. For a description of the "conventional development theory" of those years, see David Goldsworthy, *Analysing Theories of Development*, Working paper No. 12 (Melbourne, Australia: Centre of Southeast Asian Studies, Monash University, 1977).

5. See Kenneth H. Parsons, Raymond J. Penn, and Philip Raup, eds., *Land Tenure* (Madison, Wis.: University of Wisconsin Press, 1955) for the report on the Madison conference. Though not formally established until 1962, the Land Tenure Center of the University of Wisconsin was among the conference's outgrowths.

6. Raup, op. cit., p. 572.

7. United Nations Department of Economic and Social Affairs, *The United Nations Development Decade, Proposals for Action*, Report of the Secretary-General, E/3613 (1962), p. vii.

8. Ibid., p. 38.

9. Alliance for Progress, Official Documents, Pan American Union, Washington, D.C. (1961).

10. Raup, op. cit., p. 572.

11. Robert J. Alexander, *Agrarian Reform in Latin America* (New York: Macmillan, 1974), pp. 41-42.

12. For a discussion of the difficulties of administering program aid and, by inference, some of the difficulties encountered by the Alliance, see Albert O. Hirschman and Richard M. Bird, "Foreign Aid: A Critique and a Proposal," reprinted in Albert O. Hirschman, *A Bias for Hope: Essays on Development and Latin America* (New Haven, Conn.: Yale University Press, 1971), pp. 197-224.

13. Venezuela, Chile, Peru, Ecuador, and Colombia introduced land reform legislation but change was minimal.

14. United Nations, *Report of the World Land Reform Conference* (1968), p. 3.

15. Ibid., p. 66.

16. Ibid., p. 67.

17. Ibid., p. 72.

18. United Nations, *Resolutions adopted by the General Assembly during its Twenty-Fifth Session*, Supplement No. 28 A/8028 (1970), p. 48.

19. For an account of the emergence of the approach in the 1960s, see Charles F. Gallagher, "Rich Countries and Poor Countries" [CFG-1-'69], *AUFS Reports*, West Europe Series, Vol. IV, No. 9 (1969). For an account of the early and middle 1970s and the evolving issues, see Martin Kriesberg, *International Organizations and Agricultural Development*, U.S. Department of Agriculture, Foreign Agricultural Economic Report, No. 131 (1977), pp. 115-122.

20. For an account of the food situation at conference time and before, see Grant Cottam, "The World Food Conference" [GC-1-'74], *AUFS Reports*, West Europe Series, Vol. XI, No. 5 (1974).

21. Kriesberg, op. cit., p. 15.

22. For an analysis of the context and results of this increase, see Jon McLin, "Changing Fashions in Bilateral Aid to Agriculture," in Barbara Huddleston and Jon McLin, eds., *Political Investments in Food Production* (Bloomington, Ind.: Indiana University Press, 1979), pp. 203-220.

23. For evolving rhetoric at the World Bank, see Robert S. McNamara, "Address to the Board of Governors," International Bank for Reconstruction and Development, Washington, D.C. (1973). McNamara calls for "new forms of rural institutions and organizations that will give as much attention to promoting the inherent potential and productivity of the poor as is generally given to protecting the power of the privileged."

24. For developments on food affairs within the UN system and elsewhere after the conference, see Jon McLin, "Remember the World Food Conference?" [JM-2-'76], *AUFS Reports*, West Europe Series, Vol. XI, No. 2 (1976).

25. For a recent assessment, see World Food Council, *Current World Food Situation*, WFC/1980/6 (Rome, 1980). Or see "A Grim Year 2000" in *Newsweek* (August 4, 1980).

26. WCARRD, *Report* (July 1979).

27. WCARRD, op. cit., p. 6.

28. See Hirschman, op. cit., pp. 183-87.

29. On December 20, 1975, the United States House of Representatives passed the International Development and Food Assistance Act of 1975 (HR 9005), which required that most U.S. development aid be allocated to projects that benefited the poor of developing countries. The results have been decidedly mixed.

30. See also Richard E. Stryker, "The World Bank and Agricultural Development" in Huddleston and McLin, op. cit., pp. 221-240. Stryker points out that the World Bank has now substantially increased the proportion of its projects that fall under the rubric of rural development. The degree to which these projects help the "poorest of the poor" remains, of course, a key question.

31. The amount is growing. For starters, see Susan George, *How the Other Half Dies: The Real Reasons for World Hunger* (Montclair, N.J.: Allanheld, 1977); and Frances Moore Lappé and Joseph Collins, *Food First: Beyond the Myth of Scarcity* (New York: Houghton-Mifflin, 1977).

32. This and other quotations from the speech are taken from a copy distributed by the Tanzanian delegation at WCARRD.